Springer Praxis Books
Popular Science

W0037026

This book series presents the whole spectrum of Earth Sciences, Astronautics and Space Exploration. Practitioners will find exact science and complex engineering solutions explained scientifically correct but easy to understand. Various subseries help to differentiate between the scientific areas of Springer Praxis books and to make selected professional information accessible for you.

The Springer Praxis Popular Science series contains fascinating stories from around the world and across many different disciplines. The titles in this series are written with the educated lay reader in mind, approaching nitty-gritty science in an engaging, yet digestible way. Authored by active scholars, researchers, and industry professionals, the books herein offer far-ranging and unique perspectives, exploring realms as distant as Antarctica or as abstract as consciousness itself, as modern as the Information Age or as old as our planet Earth. The books are illustrative in their approach and feature essential mathematics only where necessary. They are a perfect read for those with a curious mind who wish to expand their understanding of the vast world of science.

Peter Moczo • Eva Rutšeková •
Jozef Kristek • Martin Galis •
Miriam Kristekova

Earthquakes

Tragic Challenges in History

 Springer

Peter Moczo
Faculty of Mathematics, Physics and
Informatics
Comenius University Bratislava
Bratislava, Slovakia

Jozef Kristek
Faculty of Mathematics, Physics and
Informatics
Comenius University Bratislava
Bratislava, Slovakia

Miriam Kristekova
Geophysical Division
Earth Science Institute,
Slovak Academy of Sciences
Bratislava, Slovakia

Eva Rutšeková
Faculty of Arts
Comenius University Bratislava
Bratislava, Slovakia

Martin Galis
Faculty of Mathematics, Physics and
Informatics
Comenius University Bratislava
Bratislava, Slovakia

Springer Praxis Books
ISSN 2626-6113   ISSN 2626-6121   (electronic)
Popular Science
ISBN 978-3-031-64706-2   ISBN 978-3-031-64707-9   (eBook)
https://doi.org/10.1007/978-3-031-64707-9

Translation from the Slovak language edition: "Zemetrasenia. Tragické výzvy v dejinách." by Peter Moczo et al., © GRADA Slovakia s.r.o. 2023. Published by GRADA. All Rights Reserved.

This Springer imprint is published by the registered company Springer Nature Switzerland AG
The registered company address is: Gewerbestrasse 11, 6330 Cham, Switzerland

If disposing of this product, please recycle the paper.

We dedicate this book to all who enjoy reading and exploring.
Especially to those who never cease to be surprised by the relationship between
humankind and the living planet.

Foreword

The book is as exciting as a crime thriller; once you pick it up you won't be able to stop reading it. This certainly doesn't just apply to people interested in science, because the book also tells a compelling story about the tragic individual fates of people affected by the earthquake and its effects on entire regions. It contains history, seismology, economics, politics, sociology, and much more.

Vienna, Austria
May 2024

Christa Hammerl

Foreword

This book recounting the challenges to understand earthquakes and handle all resulting risks will be a page-turner for a very wide range of readers, not only for specialized communities (education, engineering, decision-makers, urban planners, all kinds of natural, social, and formal sciences), but also the general public.

As a historical book, it obviously focuses on past events, how they were perceived, what was missed, what was learnt, what could be improved, …. History is not over, and considering the world population growth and its concentration in urban centres, new major disasters will certainly strike in the lifetime of readers.

I strongly recommend this book for high school and university students: they will get an exciting idea not only about seismology, but also about many other fields and side doors it opens in order to reduce the multifaceted risks.

Some of them are scientific or technological doors, unsealing new promising avenues, for instance (among many others) through the ongoing development of new sensing technologies, from very sophisticated (in-land and sea-bottom thousands of kilometres of fibre optics, satellite imagery, …) to much simpler, affordable low-cost sensors paving the way for "citizen seismology". Some other doors may lead to operational improvements in construction technology, early warning systems, social and institutional organization, intelligent and optimal use of artificial intelligence tools to manage all pieces of earthquake-related information, to succeed in reducing the foreseen dramatic death toll and economic costs. And some of these doors are still unknown, to be discovered and explored by the younger generation: do read this book and do write the next edition with new understanding and new answers to the same long-haul challenges!

Le-Monêtier-les-Bains, France   Pierre-Yves Bard
May 2024

Foreword

This book will take you to a fascinating scientific journey to the inside of our planet and beyond as well as to an enthralling voyage through time in the quest of mankind to understand earthquakes and in his struggle to overcome them.

Earthquakes are impressive phenomena because they shake the very surface of the Earth on which we stand and live. They arise from natural forces inside the Earth, but the origin and the nature of these forces have long been a mystery. Unveiling to the reader the pieces of this mystery and retracing how earthquakes have been perceived through time and have impacted the human history are the goal and ambition of this book. Although they may be very tragic, earthquakes are the necessary mark that the Earth is a living planet. One must realize that a planet without earthquakes would not host human life.

Another important aspect of earthquakes to have in mind is that it is not the Earth shaking itself which kills people, but it is the collapse of the man-made structures they live in. Today we know how to build structures which can withstand strong ground shaking. Thus, there is hope for the future.

The heart of seismology is the study of earthquakes, but seismology is also the major field in studying the Earth because the radiation of seismic waves produced during an earthquake illuminates the deep Earth and provides the most valuable data we have on the interior of our planet. The exploration of the Earth resources, from gas and petroleum to geothermal, also relies on seismic waves.

As you will discover in this book, the study of large earthquakes is a fascinating subject. The physics and mechanics involved are almost beyond belief. An earthquake like the dramatic 2023 Turkey earthquake broke the earth surface over a length approaching 400 km. The tectonic plates in contact moved along this distance throughout a large part of Turkey by an average of 5 m. During this earthquake, rupture propagated along the fault at a speed reaching at times 5 km per second. The giant 2004 Sumatra earthquake ruptured the plate boundary under sea over more than 1200 km. The great 2011 Japan earthquake was produced by the sudden plunge under Japan of an area of the Pacific plate of 450 km in length by 150 km in depth. This plunge occurred in two minutes and reached as much as 50 m in some places.

Throughout civilizations earthquakes have been interpreted as a manifestation of mysterious powers, often associated with religious beliefs and rooted in mythologies. It is surprising that the first scientific attempts at understanding the phenomenon only came in the second half of the nineteenth century. These first clues came from geological field observations conducted after a couple of major earthquakes and suggesting that the earth shaking was produced by faulting, that is, the displacement of crustal blocks along faults. But, as you will read, a basic understanding of the mechanism involved had to wait until 1906 when the destruction of a major western city—San Francisco— shocked the advanced civilized world and required a scientific explanation.

The long path from legends to science is an enthralling thread of the book which shows how each civilization dealt with the mystery of earthquakes, attributing them at times to the presence of giant animals hidden in the insides of the Earth, at other times using a more poetic description.

You will learn that the scientific characterization of seismic waves is also surprisingly recent, dating back to the nineteenth century. As instruments to record earthquakes did not yet exist at the time, it is through purely theoretical calculations that the existence of shear waves, which are the main source of shaking and damage in earthquakes, and surface waves was established by some of the greatest mathematicians and physicists of their time.

This rapid scientific progress followed by instrumental developments marked the birth of a new science: seismology. In subsequent years, seismic waves produced by earthquakes allowed us to uncover the inside of the Earth. This knowledge has been fundamental to understand the birth and the evolution of our planet, the origin of its magnetic field, or to search for the Earth resources which have been the pillar of the Industrial Revolution. But many mysteries remained. There was no explanation of where earthquakes occur and no understanding of why some of them were deep in the Earth where rocks were thought to be melted.

As it often happens in science the understanding of this was to come from a different field: the exploration of the deep ocean. In the 1950s and 1960s, extensive programmes of exploration of the oceans were launched making use of newly developed technology. One of the most intriguing data were the magnetic profiles recorded across the ocean floors, showing a near-periodic pattern of alternate positive and negative magnetic anomalies, organized in almost perfect symmetry relatively to the topographic mid-ocean ridges. Although controversial at first, these data were shown to be the imprint of the Earth magnetic field inversions and provided the proof of a phenomenon suggested before but rejected by the scientific community: the drift of continents. From then on, combining observations from all fields of Earth sciences led to the revolutionary framework of plate tectonics. Thus, large crustal earthquakes like the San Francisco or the Turkey earthquakes are the traces of the sudden frictional motion of the plates moving horizontally past each other, while the large Japan, South American, or Mexican earthquakes are produced by the frictional motion of the oceanic plates suddenly diving under the neighbouring continental plates.

While plate tectonics was revolutionizing the Earth sciences, the exploration of the Moon and the Solar System was attracting considerable attention, an exciting adventure retraced in this book. On 20 July 1969, the day of the first human landing, a seismometer was the first instrument deployed on the surface of the Moon. Subsequent Apollo missions added more seismic instruments which would operate successfully and continuously for 8 years, recording numerous moonquakes and uncovering the structure of the Moon. Seven years later seismometers were deployed on Mars beginning to unravel the mystery of the Red Planet.

Grenoble, France   Michel Bouchon
May 2024

Preface

Exceptionally tragic events in history meant not only immediate human suffering but also forever marked the further development of humanity.

This particularly applies to the most tragic earthquakes and tsunamis. It was by no means just a natural disaster and dealing with its visible material consequences. Especially since the Enlightenment era, the greatest earthquakes and tsunamis that hit inhabited areas always brought about a significant change in the approach of scientists, politicians, and the public to natural threats. They always also brought unexpected and surprising progress in our understanding of the planet we inhabit.

On 1 November 1755, Lisbon was destroyed by an earthquake and tsunami. Literally nobody understood what had happened and why it had happened. The scholars of the world realized that they did not actually know what was going on beneath the surface of the Earth and throughout its interior. And yet they already knew so much about the planets and moons of the Solar System and their motions at huge distances from the Earth. A truly physical theory of the world, *Philosophiæ Naturalis Principia Mathematica*, was published as early as 1687 by an English physicist, mathematician, astronomer, alchemist, and theologian Sir Isaac Newton (1642–1727).

So why did no one even suspect how and why earthquakes and tsunamis originate?

Simply because the interior of the Earth is, with a few negligible exceptions, inaccessible to direct observations and measurements. Earth physicists (geophysicists) are dependent on the measurements of physical quantities and other observations on the Earth's surface to literally decipher what the Earth's internal structure is and what processes take place in it. This is mathematically and physically very interesting and challenging.

Large earthquakes and tsunamis bring not only tragedies and destruction. They are also a chance to learn how to better prepare for them in the future, as they will threaten the increasingly densely and intricately populated regions of the Earth's surface.

We tried to write this book for all—from school pupils to seniors—to explain what fascinating phenomena large earthquakes and tsunamis are in the life of our planet and mainly in the lives of its inhabitants.

Because earthquakes and tsunamis are both natural and social phenomena, it is natural, and important at the same time, to look at them from two perspectives: that of a seismologist and that of a historian.

In a user-friendly way we explain what earthquakes and tsunamis are. Then we bring interesting stories—of seismology and earthquakes and tsunamis that changed the attitude of people, governments, and scientists towards our planet. We finish with seismology beyond our planet Earth—on the Moon and Mars.

Bratislava, Slovakia

May 2024

Peter Moczo

Eva Rutšeková

Jozef Kristek

Martin Galis

Miriam Kristekova

Acknowledgements

We are grateful to all who helped us and encouraged us.

We greatly acknowledge illustrations by Ladislav Csurma.

We especially thank our reviewers Dr. Pierre-Yves Bard, Dr. Michel Bouchon, Prof. Ondřej Čadek, Dr. Christa Hammerl, Prof. Roman Holec, Prof. Philippe Lognonné, and Prof. Peter Markoš.

We greatly appreciate the help of Dr. Andrej Cipciar, Prof. Susana Custódio, Mgr. Jitka Dobbersteinová, Prof. Hayrullah Karabulut, Associate Professor Leonard Kornoš, Prof. P. Martin Mai, Prof. Jozef Masarik, Prof. Diego Melgar Moctezuma, Prof. Pavel P. Povinec, Dr. Antonio Rovelli, and Prof. Juraj Tóth in the preparation of specific chapters.

Dr. Daniela Čadková and Dr. Jana Moczová also carefully read the entire text and helped us improve it.

We greatly acknowledge the help of Dr. Barbora Králičková from GRADA Publishing House and Dr. Lisa Scalone from Springer Nature.

Vladimír Kuric is acknowledged for portraits of the authors.

Contents

About the Authors

Peter Moczo is Professor of Physics at the Faculty of Mathematics, Physics, and Informatics at Comenius University in Bratislava, Slovakia. He is Member of the Learned Society of Slovakia and serves as Head of the Division of Physics of the Earth at the Department of Astronomy, Physics of the Earth, and Meteorology.

Prof. Moczo supervises the Master's programme in Physics of the Earth and the PhD programme in Theoretical and Mathematical Physics. Additionally, he is Member of the Scientific Council of Comenius University and the Presidium of the Slovak Research and Development Agency. Furthermore, he is a researcher at the Earth Science Institute of the Slovak Academy of Sciences.

He studied Physics at Charles University in Prague from 1975 to 1980, where he completed his PhD in Seismology in 1988 under the supervision of Prof. Vlastislav Červený. In 1999, he earned the highest scientific degree, DrSc. (Doctor of Sciences), in Geophysics.

Throughout his research career, Prof. Moczo has focused on the development of finite-difference modelling of seismic wave propagation and earthquake motion in rheologically and structurally complex media. He has also contributed to the numerical modelling of site effects of earthquakes. His extensive contributions include being the first author of the monograph *The Finite-Difference Modelling of Earthquake Motions: Waves and Ruptures*, published by Cambridge University Press, along with other books.

Prof. Moczo's scientific achievements have been recognized with awards from the Minister of Education and Science, the Rector of Comenius University, and the Presidium of the Slovak Academy of Sciences.

Eva Rutšeková is currently a Master's student of History at Comenius University in Bratislava, Slovakia. Her bachelor's thesis deals with the historical earthquakes in Komárno in 1763 and 1783.

In her role as a scientific editor's assistant, she made notable contributions to the anthology "On Enlightened Prime Minister, Christmas Carp, and Slovak Greta", a printed compilation of scientific articles featured in a section dedicated to science and research within a digital daily news platform.

In 2022, Eva Rutšeková achieved the remarkable feat of securing the third position in the Falling Walls Lab Slovakia, showcasing her commitment to her academic and scientific endeavours.

Jozef Kristek is Associate Professor of Physics at the Faculty of Mathematics, Physics and Informatics at Comenius University in Bratislava, Slovakia. He is

Head of the Physics Section and a member of the Scientific Council of the Faculty. He is also a researcher at the Earth Science Institute of the Slovak Academy of Sciences.

He studied Physics at Comenius University in Bratislava from 1988 to 1993, where he completed his PhD in Seismology in 2002 under the supervision of Prof. Peter Moczo. In 2021, he earned the highest scientific degree, DrSc. (Doctor of Sciences), in Geophysics.

Throughout his research career, Prof. Kristek has focused on the development of finite-difference modelling of seismic wave propagation and earthquake motion in rheologically and structurally complex media. He has also contributed to the numerical modelling of site effects of earthquakes. Prof. Kristek is the co-author of the monograph *The Finite-Difference Modelling of Earthquake Motions: Waves and Ruptures* published by Cambridge University Press.

Prof. Kristek's scientific achievements have been recognized with awards from the Minister of Education and Science.

Martin Galis is Associate Professor of Physics at the Faculty of Mathematics, Physics and Informatics at Comenius University in Bratislava, Slovakia. He is also a researcher in the Earth Science Institute of the Slovak Academy of Sciences.

He studied Physics at Comenius University in Bratislava from 1997 to 2002, where he also completed his PhD in Geophysics in 2008 under the supervision of Prof. Peter Moczo. From 2012 to 2017, he was a postdoctoral fellow in the team of Prof. Martin Mai at KAUST—King Abdullah University of Science and Technology in Saudi Arabia.

Throughout his research career, Prof. Galis has focused on numerical modelling of earthquake ruptures and seismic waves. He studied conditions at nucleation and arrest of earthquake ruptures and also variability of earthquake ground motion. Prof. Galis is the co-author of the monograph *The*

Finite-Difference Modelling of Earthquake Motions: Waves and Ruptures published by Cambridge University Press.

Prof. Galis's scientific achievements have been recognized with an award from the Minister of Education and Science.

Miriam Kristekova is a senior researcher in the Earth Science Institute of the Slovak Academy of Sciences. Furthermore, she is also a researcher and teacher at the Faculty of Mathematics, Physics and Informatics at Comenius University in Bratislava, Slovakia.

She studied Physics at Comenius University in Bratislava from 1988 to 1993, where she completed her PhD in Seismology in 2006 under the supervision of Prof. Peter Moczo.

Throughout her research career, Dr. Kristekova has focused on the time-frequency analysis of seismic signals and analyses of ambient noise and earthquake ground motion and methodology development. She contributed to the monograph *The Finite-Difference Modelling of Earthquake Motions: Waves and Ruptures* published by Cambridge University Press.

Dr. Kristekova's scientific achievements have been recognized with award from the Minister of Education and Science.

Ladislav Csurma Artist—oil paintings, acrylic, watercolour; author—comic books (*The Mystery of the Old House*, *The Great Adventures of the Little Dog*, *Pippi Longstocking*), fold-out books (*Animals of the World*, *Dogs*); illustrator (*Karate*, *Stretching*, *Pranayama*, textbooks for elementary schools); animator (series *Kuk and Bubu*, *Dada and Dodo*, *City on the Danube*, *Willy Wuhlmaus*, *Bob and Bobek*); 1990–1995—animated series *Spooky*, *Les Crobs*, *Renárt* at Studio 352 in Luxembourg; after 1995 *Tommy and Oscar*, *Winx*, *Monster Allergy*, *Huntik* at Rainbow Studio in Italy.

1

Interesting But Naive

From today's perspective we can say that people who lived near active faults always perceived the Earth's surface shaking as something surprising, strange, and ominous. However, even people who lived at greater distances from the faults could experience tremors, especially if they lived on the surface of sedimentary basins and valleys. Sediments could dramatically amplify the shaking of the Earth's surface caused by earthquakes with epicentres even hundreds of kilometres away.

In the early days of human civilizations, people had no chance of knowing about the existence of seismically active faults. And not at all about the amplifying effects of surface sediments. The first physically acceptable theory on the origin of tectonic earthquakes was published in 1910, 5 years after the publication of the special theory of relativity! The theory of lithospheric plates, which provided an explanation for major seismotectonic faults, began to take shape in the 1960s. Though the amplifying effect in a sedimentary layer was recognized after the Great Kanto earthquake in 1923, the understanding of complexity of seismic motion in sedimentary basins and valleys only emerged in the 1980s. It is no wonder, therefore, that over the centuries, various interesting and now quite naive ideas about the origin of earthquakes were developed. What is truly astonishing is the bizarre diversity and nature of many early views on earthquakes. The further we go back in time, across different regions of the world, the more we find beliefs that were shaped by various factors of varying objective or subjective value. Early ideas about the world, Earth, nature, life, humans, and animals were clearly reflected in perceptions of such an exceptional phenomenon as earthquakes, which literally shook individuals, settlements, cities, and sometimes entire regions.

© The Author(s), under exclusive license to Springer Nature Switzerland AG 2024
P. Moczo et al., *Earthquakes*, Springer Praxis Books,
https://doi.org/10.1007/978-3-031-64707-9_1

The Wrath of Gods and Unrest of the Animals

In many regions of the world, people explained earthquakes as a result of the behavior of mythical animals that were worshiped and often attributed with peculiar or supernatural qualities. In the territory of Mongolia, it was once believed that the Earth trembled when a frog moved, carrying the Earth on its back, at the command of the creator gods. People in Eastern Africa imagined that the Earth was carried on one of the horns of a cow standing on a stone placed on the back of a fish. When the cow's neck started to hurt, the cow tossed the Earth from horn to horn. The Earth trembled.

In Cascadia, native tribes interpreted earthquakes and tsunamis as a struggle between a thunderbird and a whale of disproportionate size and unprecedented strength. According to legends, all creation was on the whale's back. A thunderbird with a lake on its back circled above the whale. Storms and rain poured down from the lake on the world. The wrestling bird sank its claws into the whale's back, causing dramatic tremors. Under the pressure of the thunderbird's claws, the whale plunged with the whole world deep into the ocean. This myth is specific to Vancouver Island and northern Oregon, but its parallels are found throughout Cascadia. The thunderbird and the whale fight in paintings and engravings on Vancouver's cliff walls to this day.

The close connection with the surrounding nature and dependence on it caused that even the inhabitants of the Fertile Crescent had a keen awareness of earthquakes. Civilizations inhabiting the Mesopotamian region between the Euphrates and Tigris rivers had their own interpretation of earthquakes based on the specific conditions which they lived in.

According to the Sumerians, who inhabited southern Mesopotamia roughly from the fifth to the third millennium BC, the Earth tremors were attributed to the god of water and Earth, Ea, or to Ishkur, the god of heavy rains, vegetation, and thunder.

Probably the earliest mention of warning signs preceding an earthquake comes from the Sumerians. It is found in the legend of Inanna, the goddess of fertility, love and war. Inanna caused an earthquake by destroying a huge mountain. Before the earthquake, with Inanna's loud warning roar, rocks began to roll down the mountain and large snakes sputtered venom. This can be interpreted as strong foreshocks and the formation of surface fissures through which oil and natural gas reached the land from the bowels of the Earth. In some cases, the natural gas may have ignited, as the mention of fire outbreaks in the affected area is repeated in several versions of the myth.

The Babylonian Epic of Gilgamesh tells of three earthquakes with more than a hundred victims, which were said to be caused by a Bull of Heaven sent by the sky god Anu to the King Gilgamesh.

In Scandinavia, earthquakes were attributed to the god Loki. He was chained to a rock in an underground cave for the murder of his brother Baldr. A snake hissed above him, and venom dripped from its fangs onto his head. Loki's wife Sigyn was catching drops of venom in a bowl. As the bowl filled and Sigyn moved away to empty it, Loki threw himself from side to side trying to avoid the venom, causing the Earth to shake.

In the ideas of the original inhabitants of the territory of today's Romania, the Earth rested on three pillars – faith, hope and charity. As soon as humanity denied one of these values by its actions, the pillar weakened and the Earth shook.

China and Japan

Earthquakes have also plagued some areas of China. Earthquakes there were interpreted as the result of a disturbance in the balance of the cosmic principles of yin and yang, or the dissatisfaction of the gods with the actions of the current ruling dynasty. Yin and yang are among the celestial influences that underlie the balanced functioning of the entire universe. Yang represented the heat and dryness of the wind and sun, while yin represented the darkness and dampness of the Earth's bowels. The earthquake arose as a result of the yang energy accumulated in the yin environment. Such a state of cosmic imbalance was also said to be caused by the actions of a government that did not follow moral principles.

Japan has been dealing with frequent earthquakes from time immemorial. In some places, there are even several thousand tremors recorded by seismic stations, although they are not felt by humans. However, ideas about earthquakes were formed based on the ones that people felt, many of them causing damage. A well-known myth revolves around the giant catfish Namazu, said to be under the control of the god of thunder and sword, Takemikazuchi, also known as Kashima (named after the Shrine in the city of Kashima where he was worshiped) (Fig. 1.1).

The god was believed to restrain the restless catfish living underground or in the mud by pinning it down with a large stone called kanameishi. Supposedly, the foundation stone of the city of Kashima. When Kashima momentarily lost control over Namazu due to a lack of attention, the catfish would thrash wildly, causing an earthquake (Fig. 1.2). The myth was relatively

Fig. 1.1 According to Japanese mythology, earthquakes occur when a giant catfish, known as Namazu, starts thrashing wildly. The restless catfish is kept in check by the deity Takemikazuchi (also Kashima), who prevents it from causing havoc by pinning it down with a large stone. © Ladislav Csurma, 2023. All rights reserved

Fig. 1.2 Without supervision, Namazu causes an earthquake, which often comes with extensive fires. © Ladislav Csurma, 2023. All rights reserved

unknown, however it gained more attention after the earthquake that struck the city of Edo (now Tokyo) in 1855. For many, this earthquake confirmed the belief in Namazu, as it occurred in the tenth month when all the Japanese Shinto gods leave their shrines and gather in the city of Izumo in Shimane Prefecture. Since even the god Kashima was gone and couldn't keep the giant catfish under control, the catfish's thrashing caused a significant earthquake.

After the earthquake in 1855, illustrations of catfish, known as "namazu-e," became widespread. Traditionally, they depicted Kashima who tamed the resisting Namazu with a stone. The mythological catfish also came to symbolize an avenger of social inequality. The rich, who concentrated a great deal of wealth, had to contribute to the reconstruction of the destroyed homes and city after the earthquake.

Ancient Greeks

The territory of Greece is also a region of frequent earthquakes. The Greek god of earthquakes was Poseidon, and it was the wrath of the "Earth Shaker" that was considered to be the cause of earthquakes in Greek mythology. However, it was the views of the Greek philosophers that occupied a prominent place in the entire history of the perception of earthquakes before the emergence of seismology as a scientific discipline in the early twentieth century.

Greek philosophers were fascinated by earthquakes as a unique natural phenomenon and sought to understand it. Their rational thinking was likely not constrained by religion or survival aspects. Thales of Miletus (circa 624–546 BC), considered a pioneer of ancient natural knowledge, perceived earthquakes as a result of the Earth floating on an infinite water surface like a ship. Anaxagoras of Clazomenae (500/497–428/427 BC) believed that earthquakes occur when the level of ether inside the Earth's cavities rises and cannot escape, because the pores in the Earth are clogged with rainwater. Anaximenes of Miletus (circa 585–524 BC) hypothesized that earthquakes happen when cracks form in the Earth due to either excessive drying out or excessive saturation with water. Even Democritus of Abdera (circa 460–370 BC) connected the origin of earthquakes with water within the Earth. Heavy rains could saturate the Earth's interior to the point where the Earth cannot cope with it and reacts with movements and tremors.

None of these ideas explain why earthquakes occur in some places and not in others. They also do not account for periods of increased seismic activity, not to mention the evident logical shortcomings of each concept.

Aristotle (384–322 BC), a student of Plato, was one of the most significant philosophers of antiquity. His work encompassed virtually all areas of knowledge at the time and influenced the development of scientific knowledge for more than 2000 years. His ideas about the origin of earthquakes, which he presented in his work *Meteorologica*, were no exception. His surprisingly detailed theory was based on the concept of Earth's exhalations. Earthquakes themselves occur when a large amount of air, which has accumulated inside the Earth due to the evaporation of underground water, is released. The presence of water and evaporation are influenced by geological conditions (which cause the accumulation of groundwater inside the Earth) and by weather (which determines the presence of water as such).

Aristotle's ideas persisted essentially until the middle of the eighteenth century. A fundamental change in the view of the Earth's interior and its processes was initiated by the shocking earthquake on 1 November 1755, which, together with the accompanying tsunami, destroyed almost all of Lisbon. Following this earthquake, the first modern views on the nature of earthquakes emerged. The British naturalist John Michell (1724–1793) thought of earthquake motions in terms of Newtonian mechanics. He hypothesized that an earthquake is a vibration caused by the displacement of masses several kilometres below the Earth's surface.

It was not until 1910 and 1911 that the American geophysicist Harry Fielding Reid (1859–1944) published the first physically acceptable theory of the origin of tectonic earthquakes, based mainly on an analysis of the 18 April 1906 earthquake at the Golden Gate near San Francisco.

2

Earthquakes: A Short Introduction for Almost Everyone

We wrote this introduction to the topic of earthquakes in order to make the book understandable even without any basic knowledge of earthquakes and seismology. We do not assume either natural-science or technical education.

In the following chapters, we will use some elementary seismological terms in a generally understandable way. We believe that these chapters will be interesting for their important and often surprising information even for those who do not fully understand all seismological terms. However, if they read our short introduction, they can enjoy reading the next chapters even more.

The small earthquake glossary at the end of the book can also help any time during the reading.

Seismology

The name of seismology comes from the Greek word seismós (σεισμός), which means shaking the earth. Seismology is the part of the physics of the Earth (geophysics) that studies

- propagation of seismic waves in the Earth and seismic motion of the Earth's surface,
- structure of the Earth's, Moon's and Mars's interiors,
- earthquakes, moonquakes and marsquakes,
- detailed structure of the Earth's crust,
- seismic noise of the Earth,
- free oscillations of the Earth.

P. Moczo et al., *Earthquakes*, Springer Praxis Books,
https://doi.org/10.1007/978-3-031-64707-9_2

According to the objective of the research, the following can be distinguished in particular:

- seismology of earthquakes and earthquake hazard,
- seismology of the structure of the Earth's interior,
- prospecting seismology that makes use of the artificially generated seismic waves for finding natural resources,
- seismology of monitoring nuclear tests.

The Interior of the Earth

Fundamental to understanding the Earth's interior is that the Earth's interior is essentially inaccessible to direct measurements and observations, is large, and has a very complex material and rheological structure.

The material structure can be described as a spatial distribution of material parameter values. These values change smoothly but also discontinuously depending on the position inside the Earth. For investigating earthquakes and seismic wave propagation, the most important parameters are the incompressibility modulus, shear modulus, density and quality factors of the basic seismic wave types.

The modulus of incompressibility quantifies material's resistance to volume compression. The shear modulus quantifies material's resistance to shear deformation. Density is a material parameter well known to everyone. Finally, the inverse values of the quality factors characterize how the mechanical oscillations are damped by internal friction causing energy dissipation.

The seismic model of the Earth is a model of the spatial distribution of the values of these material parameters. It is important for understanding the propagation of seismic waves in the Earth.

The Greek Panta Rhei means "everything flows" and more loosely, or more aptly, also "everything changes" or "nothing remains the same". This is what the ancient Greek philosopher Hérakleitos (sixth-fifth century BC) had in mind. Inspired by Hérakleitos, Professors Eugene C. Bingham (1878–1945) and Markus Reiner (1886–1976) coined the term rheology to refer to the scientific discipline that studies how materials change and deform due to acting forces and stresses.

The rheological structure of the Earth can thus be understood as a spatial dependence of how the interior of the Earth deforms due to forces and stresses depending on temperature, pressure, duration of application of forces and stresses, and other physical parameters.

Three Images of the Earth's Interior

Chemical image If we could take a "chemical" picture of the Earth's interior, we would see that, very simply, the interior of the Earth is formed by a thin crust (5–10 km under the ocean, up to 80 km under continents), a thick mantle (to a depth of about 2890 km) and a core (Fig. 2.1). Again, very simply, the crust consists mainly of alkaline silicates and aluminium-silicates, the mantle of ferric and magnesium silicates, and the core of iron and nickel.

Seismic image If we use seismic waves to image the Earth's interior, we will also find that the mantle is made up of three layers of a thinner upper mantle and a thicker lower mantle, that there is a transition zone between the mantle and the core. We also find that the core consists of two parts, with the thicker outer part (about 2890–5155 km) liquid and the smaller inner part (about 5155–6378 km) solid (Fig. 2.2).

Accelerated movie Imagine if we could film the interior of the Earth continuously for several million years and then project the film very quickly. We

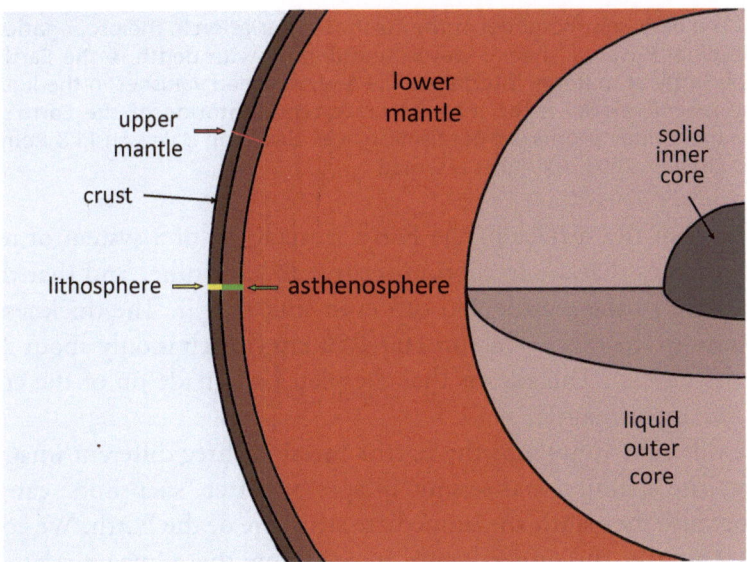

Fig. 2.1 Simplified cross-section of the Earth. The Earth is composed of a thin crust (its thickness cannot be seen in the picture), upper mantle, lower mantle, outer liquid core and solid inner core. These parts of the Earth and the depths of the interfaces were determined by the analysis of instrumental records of seismic waves. However, the properties of the interior of the Earth do not change only with depth, but also in other directions

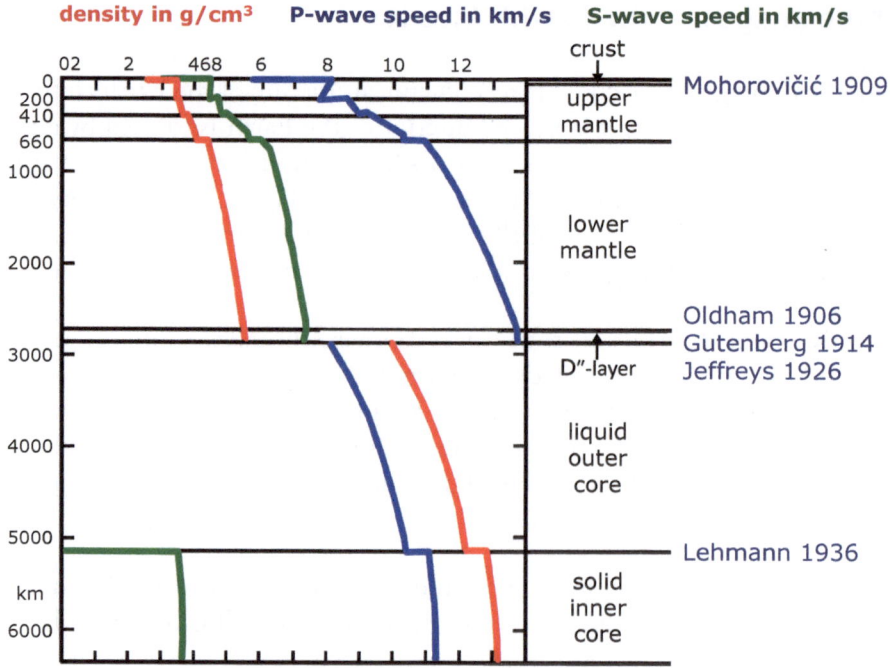

Fig. 2.2 The basic seismic model of the Earth. The variation in the propagation speeds of longitudinal P-waves, shear S-waves, and density with depth in the Earth. Values change abruptly at material interfaces. The S-wave speed vanishes in the liquid outer core. The seismic model is the most accurate representation of the Earth's interior thanks to the unique properties of seismic waves. Based on Dziewonski & Romanowicz (2007), © Elsevier, 2007. All rights reserved

would see that the surface of the Earth is made up of a system of relatively solid thin plates that are in motion relative to each other and that they also move relative to the mantle and can even sink into it. The thickness of the plates varies in the range of about 100–250 km (which is only about 2–4% of the Earth's radius). This means that the plates are made up of the crust and part of the upper mantle.

Three different views into the Earth's interior, three different images, each one true. The chemical and seismic images or better "snapshots" can be spoken of because they show the immediate structure of the Earth. We could still add, for example, the mineralogical image. How these images relate to each other, and what is happening inside the Earth, is studied by the physics of the Earth and related scientific disciplines. This is because only physics can explore the entire volume of our planet through surface, aerial and satellite measurements of physical quantities, mathematical methods and numerical modelling. The seismic model of the Earth's interior is its most accurate model.

Based on this seismic model, models of the distribution of other physical quantities and physical processes are then developed. Subsequently, other models of the Earth, such as chemical, geological and mineralogical models, are also developed.

Rocks in the Earth: Elastic Solids or Viscous Liquids?

Put simply, the rocks of the Earth's interior can behave as elastic solids or as highly viscous liquids.

Elasticity is a well-known property of many materials in everyday life. Viscosity, in fact, too. It is a measure of a liquid's resistance to deformation due to shear stresses. We can also imagine it as the "reluctance" of the liquid to flow. For example, honey has a higher viscosity than water. If we pour water into the container, it will spill immediately. If honey, it spills over the bottom of the container much more slowly.

How is it possible that one and the same rock behaves once elastically and the other time viscously? This is because the deformation of any volume of rock in the Earth does not depend only on the size and spatial distribution of the forces and stresses that cause the deformation. It also depends on the temperature and on the duration or temporal variation of forces and stresses. The fact that a given material deforms differently at different temperatures is easy to imagine. Lava expelled from a volcano flows. Volcanic rock resulting from the cooling of the lava deforms as a hard rock.

We can illustrate the dependence on duration of acting forces as follows: Imagine a sphere that is the size of a ball and is made of relatively light elastic material. The ball is thrown against the floor. The ball deforms at the time of impact and rebounds. When we catch it in our hand after the bounce, it has its original shape. However, suppose we put a weight on the ball which immediately deforms the ball as it was deformed at the time of impact and rebound from the floor. We would leave that weight on the ball for, say, a couple of years. We can imagine that if we were to remove that weight after those years, the ball would be deformed and would not have regained its original shape.

A well-known example is the behavior of corn starch mixed with water. If we prepare such a mixture, e.g., in the wash-bowl and quickly jump on it, it behaves elastically, the legs bounce off the surface. However, if we calmly stand on it, our feet will be immersed in the mixture.

This means that one and the same material can behave as an elastic solid under short-term acting force/stress and a high-viscosity liquid under long-term acting force—it slowly flows.

A short paragraph for natural-science and engineering oriented readers: there is also a very simple rheological model for this—the Maxwell model, which can be thought of as a series connection of a perfect spring and damper. Maxwell model can be characterized by a relaxation time, which is the ratio of dynamic viscosity and shear modulus. If the time of application of the forces is significantly shorter than the relaxation time of the material, the material behaves elastically. If it is significantly longer, it behaves viscously.

A material that behaves elastically at short time force application and viscously at long time force application is called viscoelastic. What is short and what is long action depends on the particular material under the given conditions. The material of the Earth's crust, mantle and inner core behaves viscoelastically from a seismological point of view. For this material, it is approximately true that the longer the time of acting forces and stresses is, the smaller the part of the Earth's interior from its surface behaves elastically. It is not easy to determine the relaxation times inside the Earth, and different values can be found in the literature.

Lithosphere and Asthenosphere

Although the Earth's crust is chemically and mineralogically distinct from the mantle, the crust and the uppermost part of the mantle form the lithosphere. The lithosphere is the well-distinguishable uppermost part of the Earth's interior in terms of dynamic processes. It is a mechanically solid layer formed by lithospheric plates. The ancient Greek word λίθος (lithos) means stone, σφαιρα (sphaira) means sphere. The oceanic lithosphere is about 100 km thick, of which the oceanic crust is about 5–10 km. The continental lithosphere is approximately 100–250 km thick, with some estimates as high as 300 km. Continental crust contributes about 30–80 km, with the thickest crust beneath the Tibetan Plateau, the Andes, and the Baltic Shield.

Asthenosphere is located beneath the lithosphere. The ancient Greek word ἀσθένης (asthénes) means weak, without firmness. The asthenosphere is a mechanically weak part of the upper mantle—the rocks in it can deform considerably without causing macroscopic fission (macroscopic cracks). It is estimated that the asthenosphere reaches a depth of up to about 420 km in some places.

The lithosphere is cold enough to be (rocky) solid. The asthenosphere is warm enough for rocks to be close to the melting point, and it is estimated that a relatively small percentage of rocks are actually melted.

Lithospheric Plates and Their Relative Motion

The radioactive decay of the isotopes of uranium ^{238}U and ^{235}U, thorium ^{232}Th, and potassium ^{40}K produces heat, which creates heat flow toward the Earth's surface. This heat and the heat accumulated at the beginning of the Earth's evolution, today primarily in the core, drive thermal convection (matter transport) in the mantle and asthenosphere. The fact that the lithosphere is broken into plates is mainly a consequence of convection acting on the relatively thin lithosphere.

Let us briefly note that the energy associated with earthquakes, volcanism and mountain formation represents only about 1% of the heat flow energy from inside the Earth's interior towards its surface.

The largest lithospheric plates with an area of more than 20 mil. km² are the Pacific, North American, Eurasian, African, Antarctic, Indo-Australian (some distinguish Australian and Indian) and South American. The smaller ones are Somali, Nazca, Sunda, Philippine, Arabian, Caribbean, Cocos, Scotia, Anatolian, New Hebrides, Amur, Okhotsk, Yangtze, Carolinian, Burmese, Juan de Fuca, ... (Fig. 2.3) There is quite a number of recognizable small plates and microplates. This is due to the complexity of the structure and dynamics of the lithosphere.

As a result of convection in the mantle, the lithospheric plates move. It should be remembered that the plates have various complex shapes and sizes.

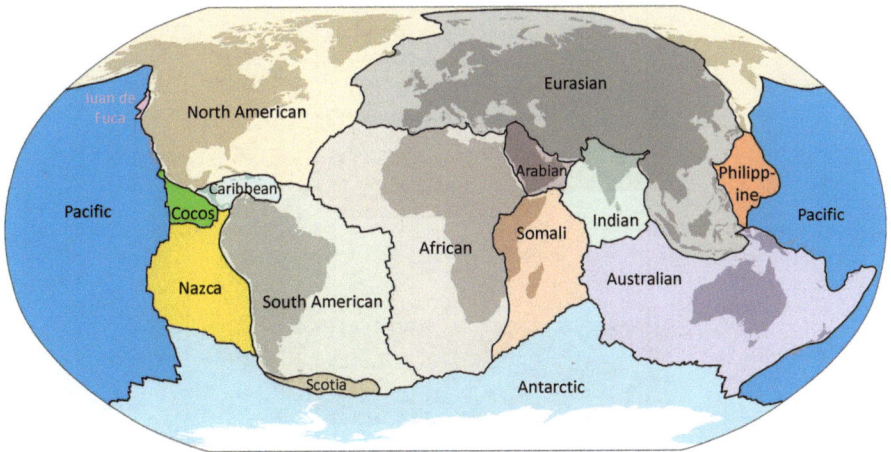

Fig. 2.3 The mutual movement of plates, into which the lithosphere is broken, is essential for the dynamics of the surface parts of the Earth. The lithosphere consists of the Earth's crust and part of the upper mantle. Modified from Wikipedia.org—public domain

The consequence is the complicated motion of one plate relative to others. This relative motion takes place mainly at plate contacts. Three basic types of contact can be distinguished: divergent, convergent, and transform (Figs. 2.4 and 2.5). We briefly characterize them, because the largest number of tectonic earthquakes occurs just as a result of the relative motion of the plates at their contacts.

Divergent Contact (or interface) On this contact, the plates move away from each other. The resulting gap is filled with material from the mantle, which moves towards the surface. This creates a new Earth's crust. The most famous example is the Mid-Atlantic Ridge, which can be seen with the naked eye in Iceland. In the Thingvellir National Park, you can literally walk in the narrow space between the North American and Eurasian lithospheric plates on fresh Earth's crust (in the time scale of lithospheric processes). Along the Mid-Atlantic Ridge, the plates are moving apart at a rate of approximately 2.5 cm/year (Fig. 2.4).

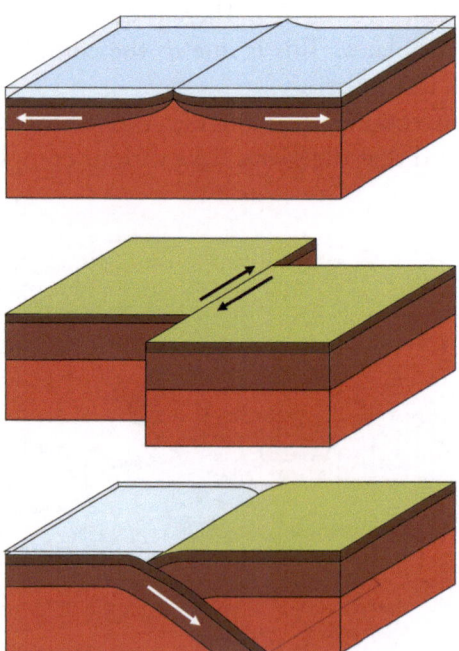

Fig. 2.4 Three basic types of lithospheric plate contacts: divergent (e.g., Mid-Atlantic ridge), transform (e.g., San Andreas fault), and convergent (e.g., dipping of the Nazca Plate beneath the South American Plate). Most earthquakes occur at convergent and transform plate contacts. Reprinted with permission from Moczo et al. (2023), © GRADA Slovakia s.r.o., 2023. All rights reserved

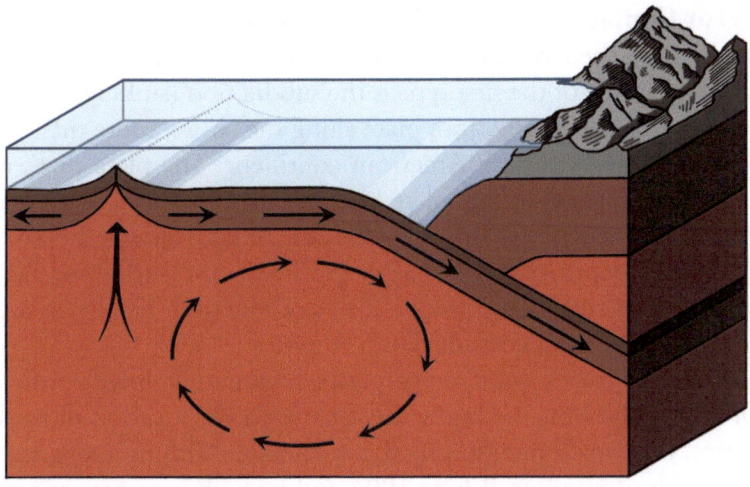

Fig. 2.5 Convection in the mantle. The upward flow leads to divergent contact, and a convection cell (indicated by arrows in the ellipse) causes movement and subduction of one plate beneath the other. The result of subduction is a deep ocean trench and a mountain range on the continent. Reprinted with permission from Moczo et al. (2023), © GRADA Slovakia s.r.o., 2023. All rights reserved

Other divergent interfaces are the Eastern Pacific Ridge, the East African Rift, the Baikal Rift Zone, the West Antarctic Rift System, and the Rio Grande Rift. Along the Eastern Pacific Ridge, near the Easter Island, the Pacific Plate is moving away from the Nazca Plate at a rate of up to 15 cm/year.

Earthquakes do not occur at the divergent interface itself. If we look at the map of the Mid-Atlantic Ridge, we see that the ridge line is broken by perpendicular faults. Earthquakes occur just on these short perpendicular faults.

Transform Contact Along the contact, the plates slide in a horizontal direction. The approximately 1300-km long San Andreas fault in California is the best known contact of its type. Adjacent to it are the Pacific and North American plates. The San Andreas fault is the so-called right-hand fault: if you stand on one plate, the other moves to the right. The plates move along the San Andreas fault at a speed of approximately 5 cm/year. A well-known right-hand fault is also the North Anatolian fault in Turkey, which separates the Eurasian Plate from the Anatolian Plate, and earthquakes at its western end significantly threaten Istanbul. The tragic earthquakes in February 2023 in Turkey occurred on the left-hand East Anatolian fault.

Convergent Contact In this case, the plates move against each other. Three types are distinguished: ocean-continental, ocean-ocean and continental-continental. An example of the first type is the subduction (sinking) of the Nazca plate beneath the South American plate along the entire, more than 6000 km long, west coast of the South American continent. The plate dives into the mantle. The subduction of one plate beneath the other caused formation of the Atacama Trench (also Peruvian-Chilean Trench) along the coast. The trench reaches a depth of more than 8 km below the surface of the Pacific Ocean. The Andes are a consequence of this subduction. Finally, so are all the volcanoes lining the west coast of South America.

In subduction zones, subducting plates retain their fragility down to a depth of about 700 km. We know this thanks to the location of earthquake hypocentres. It is worth mentioning the earthquake with moment magnitude Mw 8.3, which occured on 9 June 1994 and had a hypocentre at a depth of 636 km at a distance of about 320 km north-northeast of the city of La Paz, Bolivia. Below we will explain both the moment magnitude and the hypocentre.

An example of an ocean-ocean contact is the subduction of the Pacific Plate under the Philippine Plate. The result is the Mariana Trench reaching a depth of almost 11 km. Also, in the case of such contact, volcanoes are formed, creating island arcs.

An example of a continental-continental contact is the collision of the Indian Plate with the Eurasian Plate. The result is the Himalayas.

In fact, no contact between two large plates is simple. This can be easily seen e.g. in California. There is a surprising number of faults in the vicinity of the San Andreas fault, and it is realistic to assume that even more will be identified in the next decades. Searching for previously unknown faults is extremely important. If we don't know them, we don't know where an earthquake can be expected and what is the level of earthquake hazard in the area.

As already mentioned, most earthquakes occur at plate contacts (Fig. 2.6). The largest in subduction zones. However, earthquakes also occur inside plates—large or small. This is because even inside the plates there are blocks that are in relative motion. Interestingly, intraplate earthquakes occur mostly on or near ancient plate contacts. The Alps and the Carpathians are recent convergent zones and, for example, earthquakes in Western Bohemia originate in the site of a 350–400 million-year-old subduction zone. Intraplate earthquakes usually take significantly longer to prepare than earthquakes at the contact of lithospheric plates. Therefore, they are often very surprising.

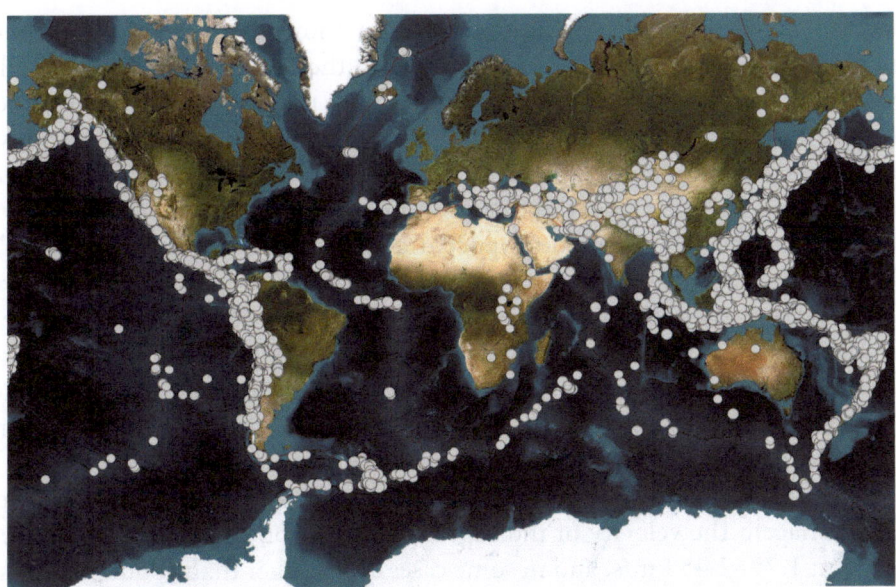

Fig. 2.6 Earthquake epicentres with a magnitude greater than 6.5 in the period 1900–2022. They can be compared with the contacts of the lithospheric plates. USGS public domain

How a Tectonic Earthquake Arises

Let's start with a simple example. A block is placed on the table, e.g., brick. A long spring is attached to the block. When we start to pull the free end of the spring, the spring starts to stretch, but the block does not move. The motion is prevented by static friction. At some point, the tension in the spring equals the static friction. Further stretching of the spring will cause the block to move.

It is relatively easiest to explain how an earthquake occurs on a transform contact, e.g., on the San Andreas fault. If we say that the Pacific plate moves relative to the North American plate along the fault, it does not mean that all those millions of years the plates simply constantly move (slide) along the fault. For example, just as two rigid blocks could slide along the contact if their contact was lubricated. Free sliding along the fault is prevented by friction.

Let's consider some point on the fault. At this point, two "particles" (or very small volumes of rock) neighbour. One particle belongs to one plate, the other particle to another plate. Due to static friction, these adjacent particles do not move—they stay together. Static friction at plate contact cannot, of course, prevent the entire giant plates from moving. If we could at one moment draw a straight line perpendicular to the fault and passing through our pair of particles, after some time we would see that the straight line around the fault in both plates was bent—deformed. Plates deform to a certain distance from the fault and tangential stress accumulates on the fault. Since the fault is not a perfect plane, the friction is not the same throughout the fault. At some point (let's call it the hypocentre and its projection on the Earth's surface the epicentre), this stress reaches the contact strength limit given by static friction. A continuing increase in stress causes a rupture (crack, loss of contact, material discontinuity): the particles "bounce" off each other—a particle of each plate to the right relative to the adjacent particle of the other plate (San Andreas fault is the right-lateral fault). Similar to the case of the brick on the table, although along the fault there will be movement of both particles.

The rupture then propagates along the contact surface (the San Andreas fault surface). The velocity of the rupture propagation usually reaches values of about 1.75–2.45 km/s, and in some cases even higher than about 3.5 km/s. For comparison, the maximum permitted speed on our highway is 130 km/h, i.e., approximately 0.036 km/s. The propagation of a rupture along a fault surface stops at a given point when the stress (that accumulated during the earthquake preparation plus that caused by the propagation of the rupture) cannot overcome the contact strength. The surface on which the originally adjacent particles lost their contact may be called the ruptured area of the fault. In small earthquakes, the size of the area may be on the order of tens of square metres. In the largest earthquakes, it can reach more than 100,000 km^2.

While the rupture propagation is a very fast and short process (depends on the dimensions of the ruptured part of the fault and can take from fractions of a second to tens of seconds), preparation of an earthquake is a very slow and long process (it can take tens to hundreds and even thousands of years, depending mainly on the rate of deformation).

As the rupture propagates, both the deformation and the stress that accumulated in the period before the earthquake occurred are being released. Due to the inhomogeneity of strain, stress and friction, usually not all of the accumulated stress is released. What remains, can lead to aftershocks that are smaller than the earthquake. Aftershocks also occur around the boundary of the ruptured part of the fault, where there is an increase in strain and stress compared to the pre-earthquake condition. Usually, in the largest of the aftershocks, about thirty times less energy is released, at most, than in the main earthquake.

The propagation of the rupture on the fault also means that even at small distances from the fault, irreversible displacements will occur. However, since the displacement (slip) at each point of the ruptured part of the fault and the displacement at small distances from the fault take place over a relatively short time (on the order of 1–10 s), nearby particles are displaced from their original positions for a short time.

Before we proceed with a tectonic earthquake, we need to explain what seismic waves are.

Seismic Waves

The interior of the Earth behaves elastically (flexibly) under short-term (fractions of seconds—hundreds of seconds) forces and pressure: if some short-acting force displaces a matter particle (a relatively small volume of rock) from its equilibrium position, the particle begins to oscillate mechanically around its equilibrium position. Elastic behavior also means that the oscillatory motion of the particle is transmitted to neighboring particles. This means that mechanical oscillatory motion propagates from the place of its origin to the surroundings. The propagation of mechanical oscillatory motion inside the Earth is called a seismic wave or a seismic wave propagation (Figs. 2.7 and 2.8).

The passage of a train or car, the impact of a heavy object, the collapse of a building, a chemical or nuclear explosion, the launch of an aircraft or a space rocket, the operation of mechanical machinery, the movement of water in a river, the surge of sea waves, the movement of magma inside a volcano, changes in air pressure, the "beats" of a person's feet when walking on a solid surface, and other short-term natural phenomena or technogenic processes

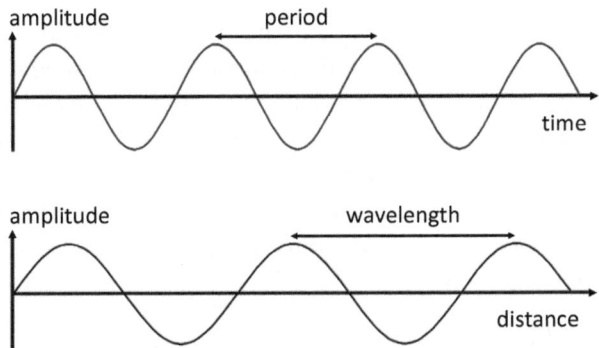

Fig. 2.7 A wave with a constant period (and hence frequency, which is the inverse of the period). Upper part, the dependence of the displacement of one particle on time; lower part, the dependence of the displacement at one time instant on location. The wavelength is equal to the product of the period and the wave propagation speed

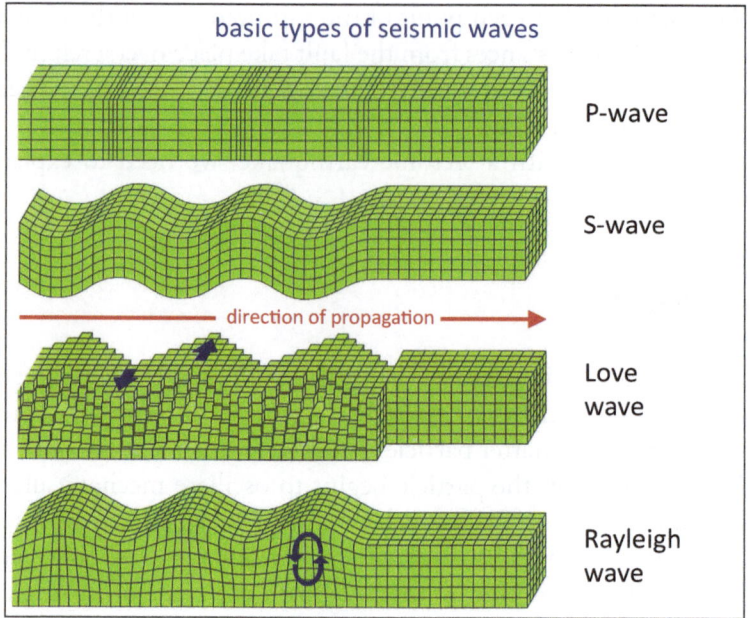

basic types of seismic waves

P-wave

S-wave

direction of propagation ⟶

Love wave

Rayleigh wave

Fig. 2.8 The figure shows how a small cube of elastic material moves and deforms as seismic waves propagate

cause mechanical oscillatory motion at a given location, which propagates to the surroundings. Thus, all of the above processes generate seismic waves. The waves can be recorded by sensitive seismometers at smaller or larger distances from the point of generation (source)—depending on how big the motion at the source is and what the transmission characteristics of the Earth's interior are between the source and the seismometer.

The simplest type of seismic wave is the so-called P-wave. As the P-wave propagates, particles vibrate in the direction of wave propagation, and only volumetric deformation of the material occurs. Before we continue, let's first explain what a plane wave is.

If we throw a pebble onto a calm surface of a pond, we know very well that circular waves spread from the point of impact, and thus, the wavefront is a circle. We can also imagine a point source in space from which a wave is radiated, and its wavefront is a sphere. If, for simplicity, the wavefront is a plane, we refer to it as a plane wave.

Let's imagine a plane wave and a cube of material with a wall perpendicular to the direction of the wave propagation before the arrival of the P-wave. The passage of the P-wave causes alternating compression and expansion of the cube in the direction of the wave propagation without changing the angles between the walls. The P-wave is therefore a longitudinal wave. Longitudinal waves can also propagate in fluids.

The second fundamental type is the so-called S-wave. When the S-wave propagates, particles vibrate perpendicular to the direction of wave propagation, making it a shear wave, and only shear deformation of the material occurs. Let's again consider a wave with a planar wavefront propagating horizontally, for example, from left to right. Let's also consider a cube of material with a wall perpendicular to the direction of the wave propagation before the arrival of the S-wave. If the S-wave is polarized in the vertical direction (particles vibrating vertically), the passage of the S-wave causes alternating deformation of the cube into a rhomboid (rhombus), with its vertical wall parallel to the direction of propagation forming a parallelogram. It is important to note that the volume of the rhomboid is always the same as the volume of the original undeformed cube at any moment. Let us emphasize that shear waves cannot propagate in fluids—fluids do not support a shear deformation. The outer core is a good example.

A longitudinal wave is faster in every medium than a shear wave. This is also where the designations P and S come from. If no other waves are propagating in the medium, faster longitudinal waves arrive at the observer's location first (Latin: Primae), followed by slower S-waves as the second ones (Latin: Secundae).

Earlier, we mentioned the factors of quality and how their inverse values quantify the degree of damping of vibrational motion. Due to the nature of vibrations, shear vibrations are more damped than longitudinal vibrations. It is also worth noting that vibrational motion at higher frequencies is more damped than at lower frequencies.

Sometimes earthquakes manifest with a peculiar strong sound. This is caused by P-waves that penetrate the atmosphere in the form of well-known sound waves.

Tectonic Earthquake

Now we can return back to the propagation of a rupture on a fault. As already mentioned, particles around the propagating rupture are temporarily displaced from their equilibrium positions. Due to these short-term displacements, they start to mechanically oscillate. This mechanical oscillation propagates into the surrounding area of the fault. Therefore, we can say that the propagating rupture generates seismic waves (the term "radiation of seismic waves" is also commonly used in seismology). The wavelengths are long enough that, even in moderate earthquakes, the waves can propagate throughout the entire volume of the Earth without being significantly damped. For a P-wave to travel through the entire Earth, it must have a period of at least 1 s, and an S-wave must have a period of at least 5 s.

The propagation velocity of longitudinal P-waves reaches more than 13 km/s in the lower mantle, approximately 6 km/s in granite near the Earth's surface. The propagation speed of shear S-waves reaches more than 7 km/s in the lower mantle, approximately 3.5 km/s in granite near the Earth's surface. Recall the image of the seismic model of the Earth above (Fig. 2.2).

When seismic waves reach the Earth's surface, they vibrate it. If it is a strong enough motion at periods that we are able to perceive, we can feel it as an earthquake (as the ancient Greeks already named the shaking of the Earth's surface). If the vibrational motion of the Earth's surface is weak or at very long periods, only sensitive seismometers will record it.

Now we can say that a tectonic earthquake involves:

- initiation and rupture propagation on a fault,
- generation of seismic waves by the propagating rupture,
- propagation of seismic waves within the Earth,
- vibrational motion of the Earth's surface.

Other Seismic Waves

Even though the propagating rupture generates P and S-waves, the intricate structure of the Earth and its surface result in the generation of other types of seismic waves within the Earth. All of them are physically more complex than P and S-waves.

Love and Rayleigh surface waves can significantly contribute to damages during earthquakes. They are waves of an interference nature (the resulting wave is formed by the combination of wave motions). Each of them has several independent modes of propagation. Each mode is characterized by a different dependence of the propagation velocity on the period and a different dependence of amplitude on depth for a given period. Simplistically, these waves can be said to propagate along the Earth's surface because the Earth's surface is a condition for their generation and propagation. This sets them apart from P and S-waves, also called body waves, as they can propagate throughout the entire volume of the Earth in all directions (except S-waves in the liquid outer core).

The depth range of surface waves depends on the mode and period. The periods of surface waves range from about 1 s to 8 min. Very simply put, the lower the frequency (i.e., the larger the period) of the mode, the greater its depth range. For example, waves with periods up to 10 s "sense" only the upper part of the continental crust, while waves with periods above 20 s reach deeper parts of the Earth's crust. Locally generated surface waves can have periods as small as 0.1 s.

During the propagation of Love waves, particles oscillate horizontally, and during the propagation of Rayleigh waves, particles execute a retrograde elliptical motion (up to a certain depth, below which it also changes). However, in areas with very soft surface layers, Rayleigh waves may be prograde in some frequency bands.

This is just a slight indication of the complexity and, indeed, the fascinating nature of the world of seismic waves within the Earth. A basic course on seismic waves would require an entire textbook, as is the case with every part of human knowledge.

The current mathematical-physical description of the propagation of body, surface, diffracted, scattered, interface, channel, leaking, head, inhomogeneous, visible, and other waves primarily originated from fundamental research and was not driven by practical applications. An interesting note is that a portion of the theory was developed by the greatest mathematicians and physicists of their time, who explored the elastic properties of the ether. This was because, for some time, many physicists believed that light was a wave motion of the ether.

Unique Properties of Seismic Waves and a Seismic Model of the Earth

Seismic waves are unique in relation to the properties of our planet. They have relatively short wavelengths, and the seismic signal recorded by a seismometer is relatively minimally distorted after passing through the Earth's interior. The amplitudes of the waves are less attenuated (compared to other geophysical phenomena), and the propagation of waves depends only on the current state of the Earth's interior. Due to these characteristics, the seismic model of the Earth is the most accurate model of the entire interior of our planet.

Let's briefly recall how the fundamental insight into the Earth's interior was obtained through seismology. In 1906, the British geologist and geophysicist Richard Dixon Oldham (1858–1936; briefly the president of the Geological

Society of London) analyzed seismographic records and realized that the relatively long travel times of P-waves throughout the Earth's interior could be explained by the existence of a relatively large core of the Earth and the lower P-wave speed in the core compared to the part of the Earth above the core. The lower wave propagation speed indicated that the material in the core is less rigid (possibly liquid) than the material between the core and the Earth's surface.

The Croatian geophysicist Andrija Mohorovičić (1857–1936) analyzed the travel times of seismic waves of a nearby earthquake in Croatia in 1909. He identified waves that preceded direct P and S-waves, interpreting them as waves propagating in a deeper layer with a higher speed along the boundary between the deeper layer and the layer above it. In this way he discovered a material boundary between the crust and the mantle. This boundary is now called the Mohorovičić discontinuity or simply the Moho (Fig. 2.9).

In 1914, the German-American seismologist Beno Gutenberg (1889–1960), a professor at the California Institute of Technology (Caltech), calculated the depth of the core-mantle boundary. In 1926, Sir Harold Jeffreys (1891–1989), a professor at the University of Cambridge, realized that the

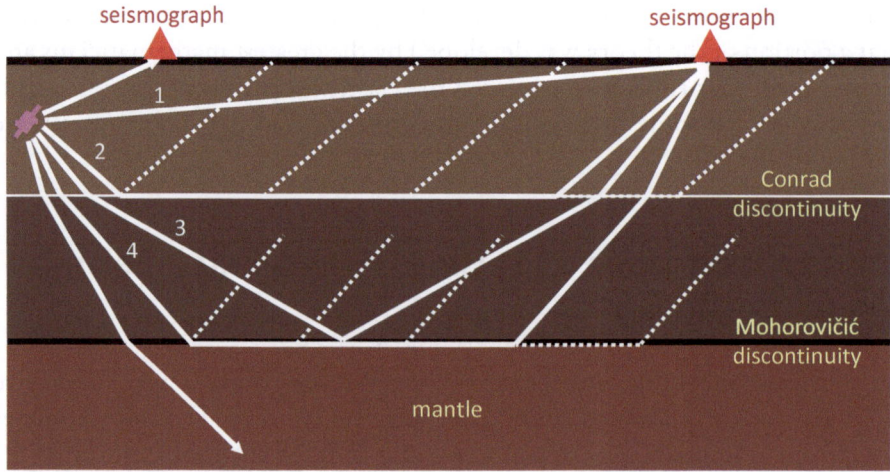

Fig. 2.9 A simplified view of the propagation of high-frequency seismic waves in the Earth's crust. The boundary between the crust and the mantle is the Mohorovičić Discontinuity. In the Earth's crust, the Conrad Discontinuity is also indicated, but it is not found everywhere. From the source to a more distant seismograph, the paths of propagation (rays) of four waves are marked: 1. direct wave, 2. wave refracted at the Conrad Discontinuity and propagating along the Conrad Discontinuity, 3. wave transmitted through the Conrad Discontinuity and reflected from the Mohorovičić Discontinuity, and 4. wave transmitted through the Conrad Discontinuity, refracted at the Mohorovičić Discontinuity and propagating along the Mohorovičić Discontinuity

absence of S-waves at epicentral distances greater than 103° could be explained by a liquid core. As mentioned earlier, shear waves cannot propagate in a liquid. Finally, the Danish seismologist Inge Lehmann (1888–1993) found that the Earth's core consists of an outer liquid core and an inner solid core. She published her discovery in 1936 (Fig. 2.10).

Today's most important method of exploring the Earth's interior is called seismic tomography. Ideally, it utilizes the complete records of earthquakes. It is called tomography by analogy with the tomography of the human body using X-rays—a well-known CT (Computer Tomography). However, in

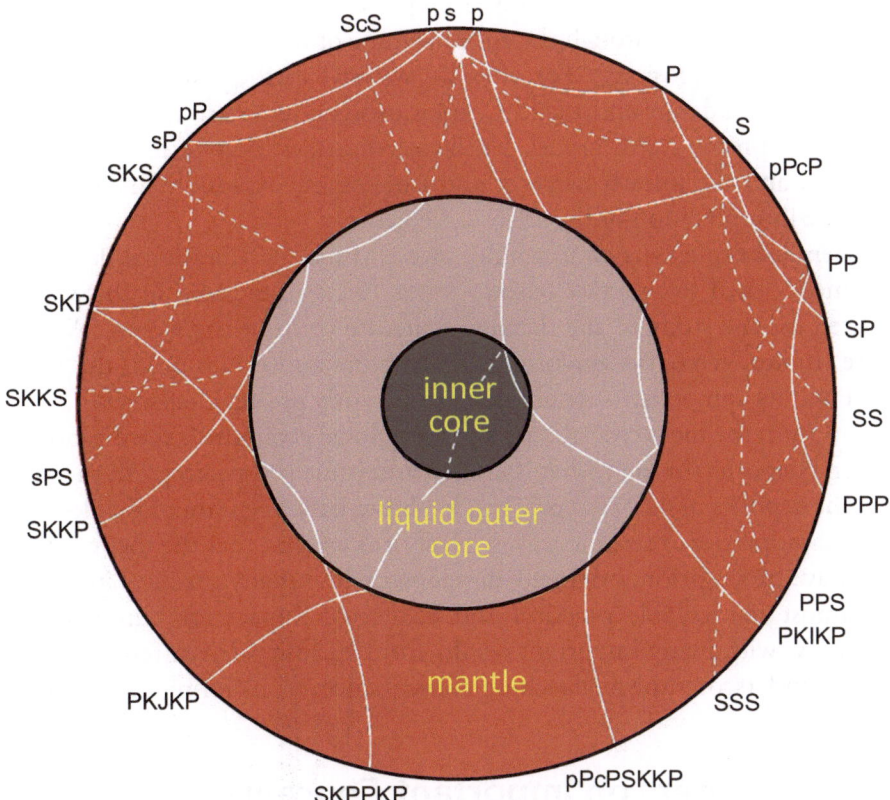

Fig. 2.10 A simplified view of the propagation of high-frequency seismic waves in the Earth. The continuous curve represents a ray of the longitudinal P-wave, dashed curve represents a ray of the shear S-wave. By tracking the rays from the source (white circle) to the Earth's surface, it is easy to determine the encoding of individual waves—reflection from the mantle-core boundary is indicated by 'c', the passage of the P-wave through the outer core 'K', the passage of the P-wave through the inner core 'I', and the passage of the S-wave through the inner core 'J', Modified with permission from Stein & Wyssession (1997), © John Wiley and Sons, 1997. All rights reserved

comparison to CT, seismic tomography is substantially more complex because seismic rays (i.e., paths of propagation of high-frequency seismic waves) are intricate three-dimensional curves. Earth is orders of magnitude larger than the human body, and it is not possible to deploy seismic stations or sources of seismic waves on the Earth's surface in a regular grid. Naturally, contrast agents cannot be used either. Therefore, seismic tomography requires more complex mathematical algorithms and significantly more powerful computers.

Completely hypothetically and hard to imagine—even if earthquakes did not occur inside the Earth, people would realize that seismic waves as a tool for exploring the Earth's interior cannot be replaced. They would begin to use artificially generated seismic waves. To replace earthquakes, which can generate waves that travel through the entire volume of the planet, they would have to use nuclear explosions. "Fortunately," we have earthquakes. However, it is true that all underground nuclear explosions conducted so far have allowed for a more precise Earth model. Unlike earthquakes, they primarily generate P-waves, and the known location and time of the explosion enable a more accurate interpretation of the seismic record.

Chemical explosions of acceptable size and mobile vibrational devices are commonly used to generate seismic waves. Waves generated in this way are utilized for the study of the detailed structure of the Earth's crust. Without these studies, geologists exploring the structure and evolution of the Earth's crust and its components would face significantly greater challenges.

However, the most crucial aspect for the global economy is the discovery of mineral deposits through what is known as seismic prospecting. All the major current deposits of oil, gas, coal, and kaolinite have been found thanks to the unique properties of seismic waves. Here, it is evident that the theory of seismic wave propagation, originally developed without any practical intent, has an almost incalculable practical and economic impact for humanity as a whole. As with many important results from curiosity-driven research in the past, it underscores the far-reaching consequences of exploring the unknown.

Short Chapters on Important Concepts

Seismic Noise

The image shows (Fig. 2.11, compare with Fig. 2.12) a seismic recording during a "quiet day" when the recording does not distinguish earthquakes, explosions, or any other transient seismic events. Nevertheless, it is clear that the

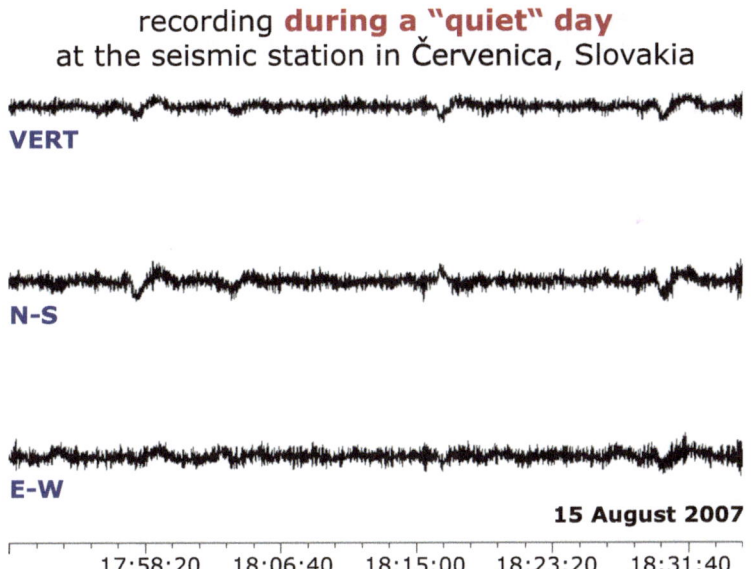

recording **during a "quiet" day**
at the seismic station in Červenica, Slovakia

Fig. 2.11 Seismometer recording at the Červenica, Slovakia seismic station. It is not possible to recognize transient seismic phenomena on the record, it is a record of continuous seismic noise. VERT, N-S and E-W indicate the vertical, north-south and east-west components of the seismic motion

Earth's surface is constantly vibrating. This is known as seismic noise (or seismic ambient noise). It is contributed to by thousands of different sources of mechanical vibration—from the weakest nearby to the strongest distant ones. These include the breaking of ocean waves, relatively distant weak earthquakes, the transmission of air pressure variations above the ocean to its floor, wind acting on trees, buildings, and hills, and all possible technical and transportation devices. Individual sources cannot be distinguished and localized in normal noise. Noise with periods less than approximately 0.5–1 s is mainly generated by human technical activities. Noise with larger periods is generated by ocean sources (wave-wave interaction among ocean waves, and interaction between ocean waves and solid Earth), on-land sources, and pressure loading deformation on land and seafloor.

Free Oscillations of the Earth

If we strike a bell with a hammer, vibrational motion propagates throughout the entire volume of the bell. After a very short time, the multiple reflections of the waves combine, and stationary vibrations of the entire bell occur in

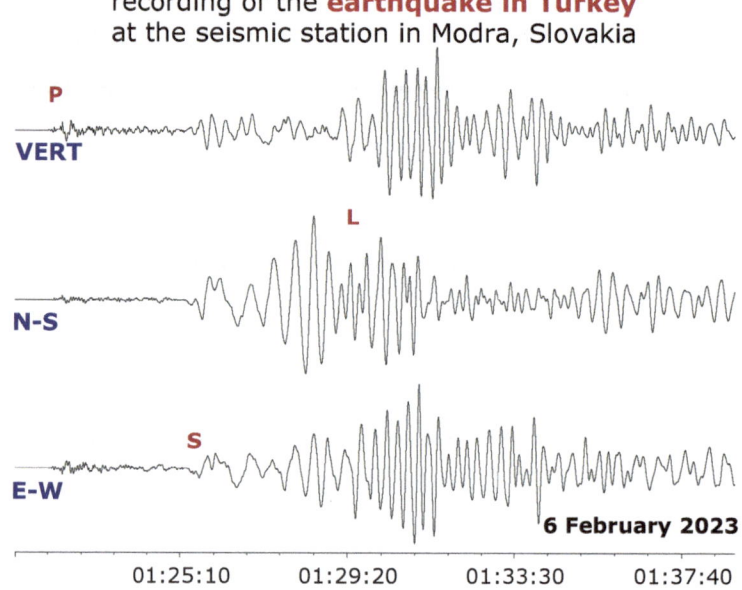

Fig. 2.12 Seismometer recording at the Modra station (Astronomical and Geophysical Observatory of the Comenius University, Modra; the station is part of the National Network of Seismic Stations managed by the Earth Science Institute of the Slovak Academy of Sciences) in the Little Carpathian Mountains. It is a record of an earthquake with epicentre in Turkey and moment magnitude 7.8. VERT, N-S and E-W indicate the vertical, north-south and east-west components of the seismic motion. The epicentre was more than 2000 km away from the Modra station. P denotes the P-wave, S denotes the S-wave, L denotes a group of the surface waves. The largest displacement at the station location was approximately 10 mm at a period of 20 s

several modes (shapes)—it is no longer possible to distinguish the propagation of vibrational motion from one place to another. We hear the sound corresponding to the frequencies of the individual modes.

During the most powerful earthquakes, a similar phenomenon occurs inside the Earth due to the superposition of surface seismic waves with long periods (tens of minutes).

In the case of the Earth, more than 1400 modes of Earth's free oscillations have been recorded and recognized by spectral analysis. Toroidal oscillations, which arise from the interference of Love surface waves, have a zero radial component, exist only in solid parts inside the Earth, do not change density and therefore neither the gravitational field. Spheroidal oscillations, which arise from the interference of Rayleigh surface waves, have all three components nonzero, exist also in the liquid core, change density and therefore also the gravitational field of the Earth. Some modes are relatively simple—for

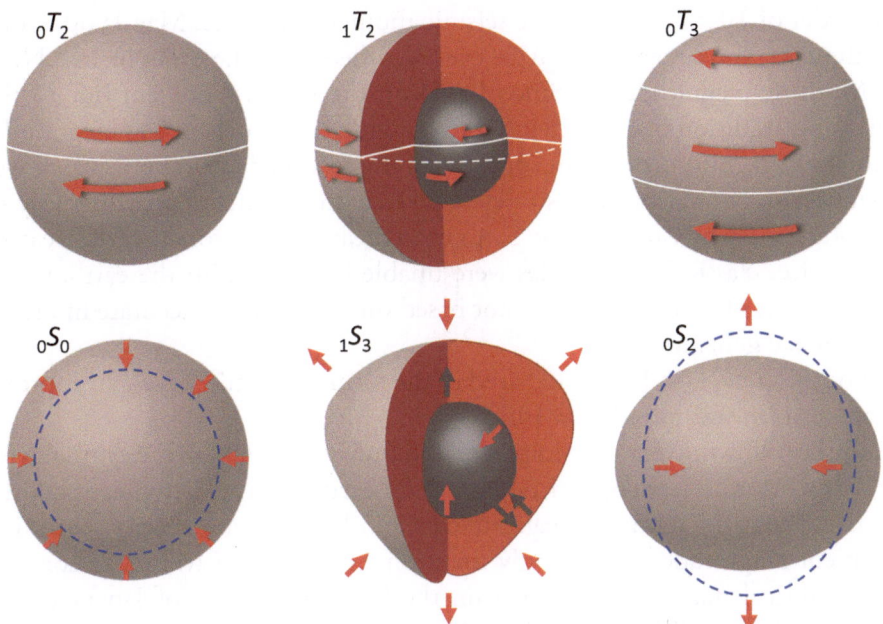

Fig. 2.13 Illustration of Earth's free oscillations. Three toroidal modes (top) can only be recorded by broadband seismometers, while three spheroidal modes can also be recorded by gravimeters, as changes in the shape of the Earth lead to variations in the gravitational field. The mode labeled $_0S_0$ represents the alternating compression and expansion of the entire Earth with a period of approximately 20 min. In the case of the earthquake on 26 December 2004, off the coast of Sumatra, the Earth's radius alternated by about 0.05 mm. The mode labeled $_0S_2$ with a period of 53.7 min is known as the 'football mode,' as it resembles a ball in the American football. Reprinted with permission from Moczo et al. (2023), © GRADA Slovakia s.r.o., 2023. All rights reserved

example, the spheroidal mode of alternating compression and expansion of the whole Earth, the toroidal mode of vibration of two hemispheres in anti-phase. Perhaps the most famous mode is the spheroidal "football mode"—the Earth deforms into the shape of a ball in American football in this mode (Fig. 2.13).

Size of an Earthquake

The magnitude of an earthquake is quantified by the scalar seismic moment $M_0 = \mu AD$, where μ is the average shear modulus, A is the ruptured area of the fault (the area over which the rupture from the hypocentre has spread), and D is the average slip on the fault. The physical dimension is Nm (Newton times metre). The seismic moment of microcracks in a laboratory rock sample is of the order of 10^{-2} Nm, while the seismic moment of the 22 May 1960, Chile earthquake, the largest earthquake ever recorded, is of the order of 10^{23} Nm.

Moment magnitude is defined based on the seismic moment due to the historical tradition of quantifying the size of an earthquake using magnitude. Moment magnitude does not suffer from the so-called saturation, which was characteristic of various types of classical magnitudes defined analogously to the original Richter magnitude. Saturation means that beyond a certain earthquake size, classical magnitudes were unable to distinguish the earthquake's size. This was because they were not based on a sufficiently accurate model of earthquake generation.

For the energy (in Joules) radiated in the form of seismic waves, an approximate relation was found $E_S = 10^{(1.5\,Mw\,+\,4.8)}$.

In the following text, for simplicity and brevity we will use the symbol Mw. If we mention, for example, Mw 9.0 for an earthquake, it will mean that the earthquake had a moment magnitude of 9.0. If we cannot specify that it is a moment magnitude, we will only mention e.g. M7.2. M will mean some of the historical magnitudes defined on the basis of the idea of Japanese seismologist Kiyoo Wadati (1902–1995) and American seismologist Charles Francis Richter (1900–1985).

Macroseismic Intensity

The effects of earthquakes on people, objects, buildings, and nature are quantified using macroseismic intensity for each location where earthquake effects have been observed. Macroseismic intensity is determined in degrees of macroseismic scale. Each intensity level is characterized by a set of features. The macroseismic intensity at the earthquake epicentre is called epicentral intensity. Unlike magnitude, which is calculated from instrumental records of seismic motion, macroseismic intensity is determined by analyzing macroseismic questionnaires. It is important to realize that one earthquake can cause different macroseismic effects in different locations. This is because these effects also depend on the distance, local geological conditions and buildings. Even intensities in the epicentral area may not simply relate to the magnitude of the earthquake.

In Europe, the most widely used macroseismic scale is the European Macroseismic Scale (EMS-98) (Table 2.1). Older scales that are still in use include the Mercalli-Cancani-Sieberg Scale (MCS), the Modified Mercalli Intensity scale (MMI), and the Medvedev-Sponheuer-Karnik scale (MSK-64). These macroseismic scales have 12 degrees. They differ in the level of detail in

Table 2.1 A brief version of the European Macroseismic Scale EMS-98

Intensity	Definition	Short description of typical observed effects
I	Not felt	Not felt.
II	Scarcely felt	Felt only by very few individual people at rest in houses.
III	Weak	Felt indoors by a few people (0–20%). People at rest feel a swaying or light trembling.
IV	Largely observed	Felt indoors by many people (10–60%), outdoors by very few. A few people are awakened. Windows, doors and dishes rattle.
V	Strong	Felt indoors by most (50–100%), outdoors by few. Many sleeping people awake. A few are frightened. Buildings tremble throughout. Hanging objects swing considerably. Small objects are shifted. Doors and windows swing, open or shut.
VI	Slightly damaging	Many people are frightened and run outdoors. Some objects fall. Many houses suffer slight non-structural damage like hair-line cracks and fall of small pieces of plaster.
VII	Damaging	Most people are frightened and run outdoors. Furniture is shifted and objects fall from shelves in large numbers. Many well built ordinary buildings suffer moderate damage: small cracks in walls, fall of plaster, parts of chimneys fall down; older buildings may show large cracks in walls and failure of fill-in walls.
VIII	Heavily damaging	Many people find it difficult to stand. Many houses have large cracks in walls. A few well built ordinary buildings show serious failure of walls, while weak older structures may collapse.
IX	Destructive	General panic. Many weak constructions collapse. Even well built ordinary buildings show very heavy damage: serious failure of walls and partial structural failure.
X	Very destructive	Many ordinary well built buildings collapse.
XI	Devastating	Most ordinary well built buildings collapse, even some with good earthquake resistant design are destroyed.
XII	Completely devastating	Almost all buildings are destroyed.

describing the effects of earthquakes on people and objects, the method of statistical evaluation of macroseismic observations, and the classification of buildings according to vulnerability and extent of damage.

In Japan, the macroseismic scale used is the Japan Meteorological Agency scale (JMA), which currently has 10 degrees (in the previous 0–7-degree scale, degrees 5 and 6 were split into 5-, 5+, 6- and 6+, degree 7 remained the maximum).

Amplified Seismic Motions

The most significant damage during earthquakes most often occurs due to seismic motion amplified and prolongated due to local geological conditions. Characteristics of such motion, for example, a maximum displacement, maximum particle velocity, maximum acceleration, duration of motion with large amplitudes, and other characteristics that can be calculated from earthquake records, can reach relatively large values due to wave phenomena in near-surface local sedimentary or topographic structures. Most of the world population is concentrated atop sediment-filled basins and valleys. Here, for basic orientation, we mention only 4 simplified models of sedimentary structures, which, however, well approximate the important effects that can occur in the real Earth (Fig. 2.14).

The sections in this chapter are written for everyone, but if some readers find them too technical, they can skip them. In the chapters on Mexico and Rome, interesting local anomalous motions will be discussed.

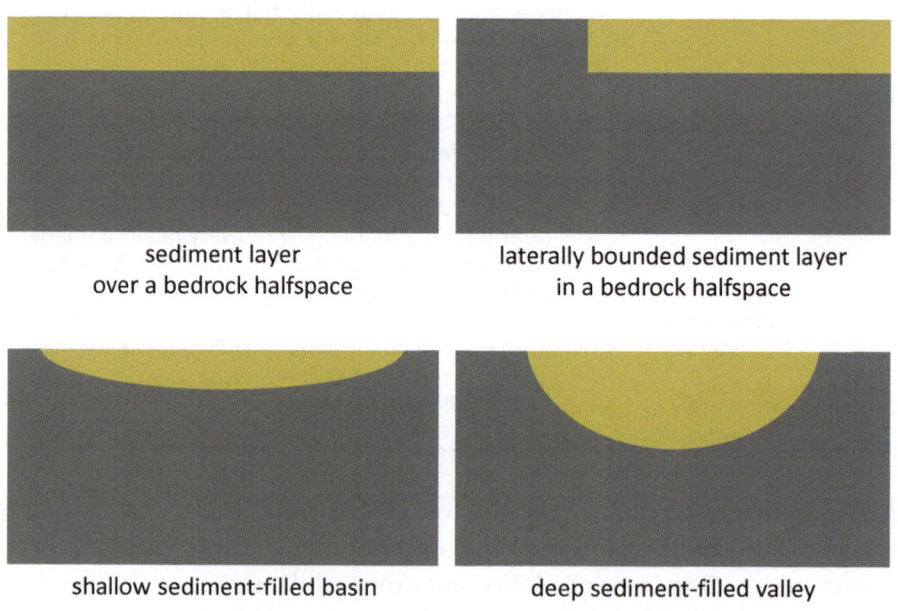

sediment layer
over a bedrock halfspace

laterally bounded sediment layer
in a bedrock halfspace

shallow sediment-filled basin

deep sediment-filled valley

Fig. 2.14 Basic simplified models of surface sedimentary structures, which are most often responsible for the greatest damage in earthquakes. Sediments are seismically less rigid than the rocky basement. They are a "trap" for waves—waves easily enter the sediments, but only a fraction of their energy gets back to the basement after reflection from the Earth's surface. Multiple reflections of waves between the surface and the lower boundary of the sediments lead to interference of waves: increasing the amplitude and prolonging the duration especially at resonant frequencies

Let's imagine a plane wave propagating vertically upward from a rocky basement. Now, it is important to consider what the wave meets near the Earth's surface. The figure shows vertical cross-sections of four fundamental sedimentary structures.

If the wave enters a horizontal layer, it subsequently reaches the surface of the sediments (i.e., the Earth's surface) and is reflected downward. Some of the energy passes through the interface between sediments and the basement into the basement, but a larger portion reflects upward, reaches again the Earth's surface, and reflects downward again. This process continues until the wave motion is completely attenuated. Waves reflected from the Earth's surface and waves reflected from the interface between sediments and the basement interfere. The waves with the biggest resulting amplitude are those with a wavelength equal to four times the thickness of the layer—resonant amplification of seismic motion occurs for this wavelength. We refer to this as vertical resonance within the sediment layer.

If the layer is bounded from the side, in addition to vertical resonance, surface waves are also generated, which propagate horizontally rightward from the vertical edge of the layer. Surface waves are formed by the superposition of waves that are generated by diffraction at the tips of the vertical edge of the layer. The formation of surface waves and their interference with the primary body waves leads to a very pronounced spatial variability of seismic motion in the vicinity of the vertical edge. For physicists—to strong differential motion.

When a wave enters a shallow sedimentary basin, something similar to a horizontal layer occurs in its center. At the edges, however, surface waves are formed by the complex superposition of waves reflected between the Earth's surface and the oblique interface between the sediments and the basement, which propagate horizontally from both ends of the basin towards the center of the basin. In the center of the basin, they meet with vertical resonance and the resulting motion can be even stronger than in the case of a horizontal layer and can last much longer.

If a wave enters a deep valley, vertical resonance in the center of the valley immediately superimposes with surface waves generated on the slopes of the valley. The result is the stationary vibration of the entire valley as a whole—literally the "natural vibrations" of the valley. This is referred to as the global resonance of the valley. Amplitudes and duration are anomalous at certain wavelengths compared to the situation where the Earth's surface is just the surface of the rocky basement.

The Number of Earthquakes

It is very difficult to estimate the total number of tectonic earthquakes in one year that could be, in principle, recorded by sensitive seismometers. This is because seismometers do not cover the entire Earth's surface sufficiently uniformly and densely. Understandably, it is much easier to find a yearly average of sufficiently large earthquakes. Based on observations since 1900 we can say that in average one earthquake with magnitude 8 and higher occurs in one year. It is interesting and for many also very surprising that such an earthquake can release more energy than all other earthquakes (earthquakes with smaller magnitudes) in one year! The biggest earthquake in one year can release energy larger than $3\,300 \times 10^{14}$ Joules, approximately more than 55 % of the total energy released by all tectonic earthquakes in one year. It is well known that even earthquakes with magnitudes between 6 and 6.9 can cause tragedy and significant damage if they hit populated areas. Approximately more than 120 such earthquakes occur in one year. Fortunately, most of them do not hit the populated areas. This is also true for approximately more than 15 earthquakes with magnitudes between 7 and 7.9.

Interesting Secondary Phenomena Due to Earthquakes

Sand Liquefaction An interesting phenomenon during an earthquake is the liquefaction of water-saturated surface sediments. This is the loss of strength in sediments due to strong seismic (vibrational) motion. If liquefaction occurs beneath buildings or other structures, it usually leads to their significant damage or collapse. The most well-known example, captured in photographs, is the partial submergence and large tilting of entire four-story residential houses during the earthquake on 16 June 1964, in Niigata, Japan.

For simplicity, let's imagine a glass container in the shape of a rectangular prism suitable for indoor aquarium. We fill the container with loose sand and slowly pour in an amount of water so that the water completely 'soaks' into the sand. In this state, the sand grains are in contact with each other, and the water fills the space between them (pores). We can talk about water-saturated sandy sediments, which are often found in many places near the seashore, as well as near rivers and lakes. The mixture is statically stable enough that we can place a heavier object on it. Then, we start vibrating the entire container with a 'suitable' frequency. The appropriate frequency depends on the specific content of the container and its shape. With the 'suitable frequency,' if the sand is rather loose, the water pressure in the pores increases to the point where the sand grains separate from each other and freely 'float' in the water. As a result, the entire mixture behaves like a liquid, and the heavy object begins to sink into the mixture. When we stop vibrating the container, the mixture will solidify again, and the object will stop sinking.

Landslides Finally, let us mention landslides. These occur quite frequently during earthquakes. If a slope is in a state close to a landslide, the vibrational motion caused by the arrival of seismic waves can destabilize it and the slope will slide.

There are other interesting phenomena. Some will be mentioned for specific earthquakes. For example, sand liquefaction and island disappearance in Chap. 4 on the origins of modern seismology. Landslides and landscape changes will be mentioned in Chap. 6 on the largest documented earthquake in Chile in 1960, anomalous seismic motions in Chaps. 7 and 8 on Mexico and Rome, flashes, sounds, radon release and methane release in Chap. 10 on the Turkish earthquakes.

3

Lisbon, 1 November 1755: The Awakening

Baixa Pombalina

From the triumphal Arco da Rua Augusta, visitors of Lisbon can admire the beautiful and elegant Baixa Pombalina and Rua Augusta with its distinct cobblestones. But Baixa Pombalina is not only beautiful. It is unique in the world for its origins.

The Great Lisbon Earthquake and Tsunami

It was Saturday, 1 November 1755. All Saints' Day suddenly turned into an unprecedented natural disaster in Lisbon approximately 20 min before ten in the morning. The biggest earthquake in the modern history of Europe caused several minutes of horror, during which houses collapsed, cracks formed in the ground and a large-scale fire started almost immediately.

A large part of the population of Lisbon was surprised by the main tremor when they were at Mass in churches. Many vaults and roofs collapsed, and thousands of burning candles contributed to the fire. Frightened people ran to an open area near the coast. They found more space than they expected as the water in the Tagus river estuary receded by several hundred metres. However, their relief and surprise at the sight of the ships on the solid bottom were soon replaced by a horror even greater than that from which they had fled. The sea returned as a tsunami more than 10 m high. Three high crests of the tsunami wave brought unimaginable masses of water to the city. About

© The Author(s), under exclusive license to Springer Nature Switzerland AG 2024
P. Moczo et al., *Earthquakes*, Springer Praxis Books,
https://doi.org/10.1007/978-3-031-64707-9_3

40 min after the main tremor, the center of the city was flooded (Figs. 3.1, 3.2, 3.3 and 3.4).

Many of those who escaped the fire and the falling debris of collapsing houses perished in the swift and huge tsunami.

Fig. 3.1 Lisbon on Saturday, 1 November 1755, at half past nine in the morning. All Saints' Day, a quiet morning, many people were at Mass in the churches or in the kitchens preparing lunch. © Ladislav Csurma, 2023. All rights reserved

Fig. 3.2 About 20 min before 10 AM, the largest earthquake in Europe's documented history shook Lisbon and much of Europe. Perhaps as many as 15,000 people lost their lives. Soon the sea receded several hundred metres. © Ladislav Csurma, 2023. All rights reserved

Fig. 3.3 The earthquake almost immediately caused a fire that spread very quickly. The situation was exacerbated a few minutes later by a 10 m high tsunami. © Ladislav Csurma, 2023. All rights reserved

Fig. 3.4 Unimaginable masses of water flooded the crumbling and burning Lisbon. Many of those who survived the violent tremors and fire perished in the devastating torrent of water. In all, up to 40,000 people may have lost their lives. © Ladislav Csurma, 2023. All rights reserved

When the waters finally receded, the fire raged again, destroying for the next five days what little had survived the earthquake and tsunami.

According to estimates, the earthquake could directly kill 10,000–15,000 residents of Lisbon. The fire and tsunami raised the death toll to perhaps as many as 40,000 people, roughly a fifth of Lisbon's total population. Some estimate an even significantly higher number of victims.

About 85% of Lisbon's buildings were destroyed. The center of the city was practically completely destroyed as a result of tremors, liquefaction under the buildings, and tsunami. Libraries, palaces, the newly built opera house, royal archives with Vasco da Gama's historical records, 70,000 volumes in the royal library and paintings by Titian, Rubens and Corregio were destroyed.

A New Era in Interest in the Planet We Live on

Terrible, overwhelming and incomprehensible. Such was the Lisbon disaster. Nobody—literally—had any idea or understanding of what had actually happened.

The Great Lisbon earthquake and the tsunami it triggered caused damage not only in Lisbon. The earthquake was also strongly felt in Germany and even in Cape Verde. Recent research indicates damage caused by the Lisbon tsunami as far away as Brazil.

We hope not to offend with an amusing remark. Church authorities immediately expressed their belief that the disaster was the consequence of God's wrath. With all due respect, this assessment did not convince many: the local red-light district remained virtually unscathed.

But let's get back to the essential aspect. Even the most educated scholars of that time did not understand what an earthquake is, where and how it originates. Practically, they knew no more than the ancient Greeks. An earthquake was simply the shaking of the Earth's surface, buildings and objects. Scholars had no idea where, how and why tsunami occurs. That's also why they couldn't even imagine why the previously unimaginable shocks and rushes of water suddenly appeared together.

Scholars were left in awe and with a depressing sense of inability to comprehend the unique and unprecedented event. More than ever, they had to realize that they had no idea what was happening inside the Earth.

However, as is often the case in such situations in the history of humanity, the Lisbon disaster marked the beginning of a new era in the interest in scientific understanding of our planet. The interest of scholars in the incomprehensible catastrophe was a real milestone—natural scientists, mathematicians, physicists, and philosophers began to scientifically study the interior of our planet and the processes inside it.

The Earthquake That Shook the Minds of the Greats—Voltaire, Rousseau and Kant

Europe, of course, knew about earthquakes. Just a few decades earlier, they had devastated Jamaica and Sicily. However, since these were relatively distant regions in the European mindset of the time, neither scholars nor the public perceived them dramatically. The greatest European philosophers, thinkers and theologians were also aware of other natural phenomena. Their symbolic value was attractive to several schools of thought.

In the case of Lisbon, however, it was an unprecedented disaster in size. Reportedly the most tragic destruction of a city at the height of its prosperity since the days of the biblical Sodom and Gomorrah. Lisbon was the fourth largest city in Western Europe. Moreover, one of the most zealous in Christian piety.

Zealous Catholics, such as the Jesuit missionary Gabriel Malagrida (1689–1761), therefore saw tragedy as the result of God's wrath for human immorality and sin. On the contrary, Protestants saw the source of misfortune in the barbaric Inquisition and fanatical Catholicism.

However, the philosophers of the Enlightenment in particular, were the most affected by the Lisbon disaster. It was they who, for the first time questioned the basic tenets of the Christian faith and Church, and reassessed their relationship with a modernizing society and the emerging experimental sciences. A large part of the European Enlightenment intelligentsia adopted the concept of deism, which provided a rational and constructive explanation for the existence of God as an abstract principle that was responsible for the creation of the world, but after its creation no longer interfered with its spontaneous evolution.

François-Marie Arouet (1694–1778), the French writer and philosopher known as Voltaire (Fig. 3.5), was staying at his Swiss residence Les Délices at the time of the earthquake. The overwhelming tragedy motivated Voltaire to write the sharply worded *Poème sur le désastre de Lisbonne, ou Examen de cet axiome: Tout est bien* (*Poem on the Lisbon Disaster Or an Inquiry into the Axiom, "All is Well"*), which became the preface to his later work *Candide, ou L'Optimisme* (*Candide: or, All for the Best*) in 1759. These works were Voltaire's response to the attitudes of the representatives of so-called philosophical optimism, who approached the Lisbon tragedy in line with their belief that we live in „the best of all possible worlds". Optimists such as Gottfried Wilhelm Leibniz (1646–1716) and Alexander Pope (1688–1744) argued that if the God is perfectly good and all-powerful, so must be his work. Evil existed in

Fig. 3.5 François-Marie Arouet (1694–1778), best known as Voltaire, was a great French writer and Enlightenment philosopher. The tragedy of Lisbon inspired him to write one of the most important philosophical works of the Enlightenment. Jean-Jacques Rousseau (1712–1778) was a French philosopher and writer. With his views he influenced the Enlightenment. In the context of the Lisbon tragedy, he highlighted a problem that is still very relevant today—the high concentration of population in megacities in earthquake-prone areas.

the world, but there was still less of it than of good. Thus, the world had to be the best of all possible worlds. At the same time, from the deistic point of view, the world functioned according to God's intention, but without his intervention. Thus, for example, in the case of a devastating natural disaster, they would've argued that the general good justifies the partial evils that occur in the world on a daily basis.

Voltaire criticized the representatives of philosophical optimism for their attempt to rationalize unnecessary evil and suffering. In his view, this was empty and inhuman philosophizing. The Lisbon disaster only reinforced his pessimistic and fatalistic view.

Voltaire received a reaction to *Poem on the Lisbon Disaster* from the essayist and social critic Jean-Jacques Rousseau (1712–1778) (Fig. 3.5). The 43-year-old French philosopher felt compelled to confront Voltaire and his exaggerated pessimism. According to Rousseau, Voltaire vehemently denounced Leibniz and Pope for diminishing the significance of human suffering by optimistically ascribing a purpose to it. But isn't his exaggerated pointing out the bleakness of life merely an unnecessary amplification of already existing suffering?

The letter that Rousseau wrote to Voltaire marked the beginning of his "return to nature". For the catastrophe caused by the Great Lisbon earthquake neither nature nor God was responsible. Man himself was to blame for his misfortune, because he built densely populated areas and in times of danger he refused to abandon his material possessions.

From a certain point of view, it can be said that Rousseau, with the foresight of more than 250 years pointed to a great humanitarian problem of our time. Today, people living in many megacities in less developed countries are indeed very threatened by earthquakes. There are several reasons for this. These include ignoring expert opinion and poor economic conditions.

Voltaire thanked Rousseau for his letter but did not respond to its content. However, Rousseau later believed that Voltaire's *Candide* was the real response.

Rousseau found a follower in the young Prussian teacher Immanuel Kant (1724–1804) (Fig. 3.6), who became one of the most determinant European

Fig. 3.6 Immanuel Kant (1724–1804) was a German philosopher and thinker, a prominent representative of the Enlightenment. Already a year after the earthquake, he published three essays in which he tried to analyze the causes of the unprecedented natural disaster. John Michell (1724–1793) was an English natural philosopher who made important contributions to astronomy, geology and physics. He was probably the first to present the view of the existence of black holes. He was the first to think of earthquakes in terms of Newtonian mechanics. He can be considered one of the first seismologists. © Ladislav Csurma, 2023. All rights reserved

thinkers and one of the main representatives of the Enlightenment. The Lisbon catastrophe fundamentally changed the worldview and philosophical outlook of the then 32-year-old Kant. Kant had published three essays on his theory of the origin of earthquakes in 1756. Kant noticed, for example, the connection between earthquakes and the outline of mountains. He also thought that the origin of earthquakes was related to the mechanism of the great fire inside the Earth, by which nature possessed hot springs, fertile soil and iron. The tremors themselves were the result of underwater cavities caving in.

Nor did Kant omit to criticize the man who in his vanity built huge cities. The evil that caused the catastrophe could not have been a form of divine punishment—man himself was, after all, a small part of nature, and attributing himself greater value would be egotistic. He realised that the Earth and its processes were indifferent to man.

The eminent English naturalist John Michell (1724–1793) (Fig. 3.6) should also be mentioned here. Michell contributed new knowledge to physics (especially astronomy, optics and seismology) and geology. Michell was the first to hypothesize the existence of black holes. After the Great Lisbon earthquake, he thought of earthquake motions in terms of Newtonian mechanics. He hypothesized that an earthquake is a wave motion caused by the displacement of masses a few kilometres below the surface of the Earth.

The Great Lisbon earthquake, whatever tragic, whatever natural, played a very significant role in the formulation of great ideas and in the creation of works which, in their timelessness, reach the thoughtful readers of today. The very discussion of it was an 'earthquake' that shook many ideas and ideals that had been taken for granted until then.

Portugal's Colonial Ambitions

With the rise of the absolutist monarchy in the early eighteenth century, Portugal began to grow richly. "The Portuguese Sun King" João V (reigned 1706–1750) was inspired by the absolutist rule of King Louis XIV of France, and longed for a royal court of wealth and extravagance that would match that of France. In doing so, he attempted to make Lisbon a second Rome. The Kingdom had the favour of the Roman Catholic Pope, a commercial treaty with England, a geographical location that allowed to keep out of European conflicts and a seemingly endless gold reserves in far-off colonial Brazil.

Brazil was literally a gold mine for the Portuguese, mainly after they lost some of their colonial possessions in India as a result of colonial wars with the

English, French and Dutch. One-fifth of the proceeds of Brazilian gold went directly to the royal treasury. Brazilian gold accounted for about 80% of all gold held by European countries in the eighteenth century. Even as the royal family grew richer, masses of ordinary people emigrated to Brazil, mainly for the opportunity to acquire a land with slaves in the hope of a better life.

The Great Lisbon Earthquake significantly worsened the economic situation and contributed to Portugal's decline as a colonial power. The enormous material damage caused by the earthquake, tsunami and fire, the financial consequences of the trade treaty with England, which ultimately proved disadvantageous, and the colonial predation of the European colonial powers forced the Portuguese to concentrate on the reconstruction of the ruined capital.

An Enlightened Prime Minister and the Reconstruction of a Destroyed Lisbon

King José I of Portugal (reigned 1750–1777) and the whole royal family only escaped death with good fortune. Survival of the royal family, and thus the preservation of formal political stability, was a good prerequisite for the rapid reconstruction of the ruined city. Prerequisites are a necessary condition. But not sufficient. As always, the situation required exceptional creative and coordinating personality. It emerged that this personality would be Sebastião José de Carvalho e Melo (1699–1782), known as the Marquis of Pombal since 1769 (Fig. 3.7). It is significant that he was the Secretary of State of Foreign Affairs and War from 1750 to 1756. In 1756 he was appointed the Secretary of State of Internal Affairs of the Kingdom, position which he occupied until 1777. Usually, Marquis of Pombal is referred to as a prime minister of Portugal or the chief minister to King José I. In this chapter we will simply call him a prime minister.

The prime minister mobilised troops, organised aid and shelter for the survivors. He immediately began planning the reconstruction of the city. He prevented the outbreak of the epidemic by having the dead buried immediately, without any religious rites.

The prime minister devised a macroseismic questionnaire, which is even today an indispensable tool in seismology for obtaining information about the effects of earthquakes. (Such information is as important as the data that seismologists obtain with seismometers.) In the questionnaire distributed in Portugal, the prime minister asked people if they had observed a change in the

Fig. 3.7 Sebastião José de Carvalho e Melo (1699–1782), known as Marquis of Pombal since 1769, was the prime minister of Portugal. An enlightened prime minister who mobilised the army, organised aid and shelter for survivors, and prevented the outbreak of an epidemic. He immediately began to plan the reconstruction of the city. He devised a macroseismic questionnaire and introduced anti-seismic requirements in the reconstruction of Lisbon.

level of water in wells, if they had observed any strange animal behaviour, and he wanted people to describe the damage to buildings. He even asked whether the movement of the Earth's surface was stronger in any direction compared to other directions. What we know today about the Great Lisbon earthquake is mainly due to the data that the prime minister obtained using the macroseismic questionnaire.

It can be said without any exaggeration that the prime minister Sebastião José de Carvalho e Melo was one of the first seismologists.

The prime minister has not only changed the geometry of the streets and buildings in the reconstruction of Lisbon. He laid down strict rules for the construction of buildings in a sober and restrained style, which is still called *estilo pombalino* after its inventor. The central part of Lisbon was built to withstand future earthquakes. The prime minister had models of the houses built and had them experimentally tested for their resistance to shaking. The shaking was taken care of by soldiers who marched around the models. In doing

Fig. 3.8 A view of Praça do Comércio and the beautiful historic centre of Baixa de Lisboa, also known as Baixa Pombalina. It was built after a devastating earthquake with the application of innovative anti-seismic measures thanks to the Marquis of Pombal. © Ladislav Csurma, 2023. All rights reserved

so, the prime minister essentially invented the precursor to today's vibration tables that structural engineers use to test the earthquake resistance of buildings.

Baixa Pombalina (Fig. 3.8) is thus not only beautiful, but also a unique part of Lisbon. The consequence of a disaster, the result of the efforts of an enlightened prime minister.

What the prime minister Sebastião José de Carvalho e Melo was able to do was probably the result of a confluence of circumstances—the need, but also the opportunity, to rebuild Lisbon, the powers of the prime minister, the Enlightenment, an understanding of the need for scientific knowledge in dealing with the aftermath of the disaster, and, last but not least, an exceptional intellect.

Seismologists' View of the Great Lisbon Earthquake

The hypocentre of the largest earthquake in the documented history of Europe was about 290 km southwest of Lisbon. Yes, under the Atlantic Ocean. Therefore, Lisbon was hit not only by an earthquake, but also by a tsunami. The estimated moment magnitude is Mw 8.5 to 9.0. For comparison, the moment magnitude of the 2011 Tōhoku earthquake in Japan, which caused a catastrophic tsunami, was Mw 9.0. The magnitude of the earthquake that caused the disaster in Turkey on 6 February 2023 was Mw 7.8.

However, what happened on 1 November 1755 is still not satisfactorily explained, although much research has been devoted to the earthquake. The attempt to investigate the origin or source of the Great Lisbon earthquake encounters fundamental problems. There are no data objectively recorded by measuring instruments, since the necessary instruments were not yet available at that time. The earthquake occurred at a time when no one understood how earthquakes occurred. The available contemporary and historical documents are obviously affected by this fact and therefore do not constitute full-fledged information in this sense. Since the hypocentre was below the Atlantic Ocean, there are no, not even lay observations in the epicentral region.

Essentially, only data on the macroseismic effects of the earthquake, i.e., estimated values of macroseismic intensity, could be used to estimate the location of the epicentre and the magnitude of the earthquake. Moreover, those had to be extrapolated, because the epicentre was located under the ocean. However, the spatial distribution of the macroseismic intensity values on land is too complex for a reliable estimation of the epicentre location, epicentral intensity and also for the subsequent estimation of the earthquake magnitude.

In addition to the lack of data, the situation is also complicated by the unique and complex tectonic situation. In the past decades, several marine geophysical cruises were carried out to find the source of the Great Lisbon Earthquake. Based on the measurements, many tectonic faults were found and mapped. A diffuse boundary of the African lithospheric plate was also detected in the area with several faults that behave like a network of faults and several of them can be activated during one earthquake. Moreover, the lithosphere in this area is among the oldest oceanic lithospheres on the entire planet. It is therefore so fragile that earthquakes often occur in it at depths of up to around 40 km. In places where one lithospheric plate is not subducting beneath another, earthquakes usually only occur at depths of 15-20 km. However, it is not known how this tectonic activity at depths of up to 40 km is related to the shallow faults in the Earth's crust that seismologists have already identified.

Precisely due to the complicated tectonic situation, it is still unknown on which tectonic fault a more recent, the so-called Portuguese earthquake of 28 February 1969, occurred. This is despite the magnitude Mw 7.8, which means it was well recorded by hundreds of seismic stations around the world.

Around the time of the Portuguese earthquake, and thus at the beginning of the theory of lithospheric plate tectonics, seismologists concluded that the

Great Lisbon Earthquake originated in the area of the undersea Gorringe ridge. Later, when tsunami arrival times were used to estimate the location of the epicentral area, it turned out that the tsunami probably originated closer to Lisbon, not as far as the undersea Gorringe ridge is.

Current research, in an attempt to correlate the estimated magnitude of the earthquake and the arrival times of the tsunami, is focused on the possibility that the Great Lisbon Earthquake is the result of rupture propagation on several faults. However, the very unreliable estimation of the magnitude value remains a problem.

A better chance of solving the problem is a numerical modelling that would take into account tsunami arrival times, macroseismic intensities, and mapped faults. This is a physically and computationally very complex challenge. However, it would be very good if seismologists could successfully tackle it.

4

From Naive Beliefs to Modern Science

How did seismology, as we know it today, evolve from the naive beliefs mentioned in the first chapter? The knowledge and tools necessary for the development of seismology as a scientific discipline had been developing largely independently along several parallel lines for a long time: 1. Observations of where and when earthquakes occur and their effects. 2. Contemplations about how and why earthquakes arise. 3. Attempts to instrumentally detect earthquakes, and measure and record tremors. 4. The development of mathematical tools and physical theories for the emission and propagation of waves in elastic solids. Only when all these lines of research converged, and instruments were made capable of recording even distant earthquakes, could seismology emerge as a distinct scientific discipline, leading to its rapid development.

Observing Earthquakes and Their Effects

Naturally, people have been observing earthquakes and their effects since ancient times. Historical records of when and where they occurred, including descriptions of their effects, can be considered the earliest ways of monitoring earthquakes.

Records of significant earthquakes, primarily based on the damage they caused, have been preserved since the times of ancient Greece and China. The oldest documented earthquake in China dates back to around 1831 BC, with a brief record simply stating that a mountain shook. More extensive records come from the period after 780 BC during the Zhou dynasty (1046–256 BC).

P. Moczo et al., *Earthquakes*, Springer Praxis Books,
https://doi.org/10.1007/978-3-031-64707-9_4

Some of these records were so detailed that modern studies have been able to use them for assessing the extent of damage and estimating the earthquake magnitude. However, the Chinese scholars' ideas about the origin of earthquakes were rooted in mythology or had other spiritual explanations. It was thinkers in ancient Greece who began to consider earthquakes as natural phenomena. They noticed, for instance, that earthquakes had more pronounced effects in areas with soft, moist, and loose substrates than in dry and rocky areas.

Although old earthquake records are valuable and irreplaceable, they are generally not suitable, with a few exceptions, for determining the size or effects of earthquakes. A significant change occurred especially after the catastrophic Lisbon earthquake in 1755.

18 November 1755, Cape Ann Earthquake

The Cape Ann, Massachusetts earthquake is considered to be the most damaging earthquake in Northeastern US history and the first earthquake in North America to have a published report written by a scientist. It was elaborated in 1758 by John Winthrop IV (1714–1779), professor of mathematics and natural philosophy at Harvard University. The surviving data allowed an estimate of the magnitude, Mw 5.9–6.7, and the maximum accelerations of the seismic motion in the Boston area. We cannot rule out that this earthquake could have been initiated by the Lisbon earthquake—as an "aftershock" at a distance. It is interesting that the earthquake was also felt by sailors on a ship approximately 350 km from the coast. After the earthquake, they saw a large number of dead fish on the surface of the sea.

Three Earthquakes Near the City of New Madrid in 1811–1812

A surprising event was the sequence of three earthquakes occurring shortly one after another in the New Madrid area in the Mississippi River Valley in the USA. The first occurred on 16 December 1811 (with an estimated magnitude of 7.5), the second on 23 January 1812 (magnitude 7.3), and the third on 7 February 1812 (magnitude 7.5). In addition to the three main shocks, there were several significant aftershocks, with 8 of them felt as far as 320 km away. Over the 3 months following the first earthquake, up to 1874 earthquakes were observed in this area.

The strongest shocks awakened people at distances greater than 1400 km, including then-U.S. President James Madison in the White House in Washington, and rang bells in Boston. Visible surface waves were observed, similar to waves on water or the movement of wheat in strong wind. At the Reelfoot Fault in Tennessee, where one of the three earthquakes occurred, one side of the fault moved down relatively to the other side by up to 6 m. The area that subsided was inundated by water from the Mississippi River, creating Reelfoot Lake. Other extensive areas were uplifted or subsided, sometimes by several metres. The banks of the Mississippi River collapsed in many places, along with trees, and in some areas, the river temporarily flowed in the opposite direction, creating temporary waterfalls and new lakes. Some islands in the river disappeared. Passengers on the first steamboat voyage on the Mississippi River found one morning that the island where they had anchored the previous evening was no longer there. During the earthquakes, significant liquefaction occurred. Water and sand erupted from cracks in the ground, creating formations resembling sand craters on an area of up to 10,400 km². Remnants of these formations are still preserved today. Landslides occurred over a large area, including mild slopes.

The earthquakes in the New Madrid area were surprising not only because there were three such large earthquakes in quick sequence, but also because they occurred within a region that was considered peaceful. Only recent research has shown that similar sequences occurred around the years 900 and 1450 as well. A detailed report analysing the preserved records and describing the consequences of the three earthquakes in 1811–1812 was published by geologist Myron L. Fuller over a century later, representing one of the first significant seismological works published within the USGS (United States Geological Survey), which was established in 1879.

The 1857 Great Neapolitan Earthquake and Robert Mallet

The earthquake in the Italian province of Basilicata, southeast of Naples, was significant both due to the destruction it caused and the subsequent research. Irish geophysicist and civil engineer Robert Mallet (1810–1881) (Fig. 4.1) conducted a thorough scientific investigation in the field, employing his engineering knowledge and physical principles.

In his request for research support from academic societies, he emphasized that only the application of a scientific approach in earthquake analysis could bring progress. Mallet received support from the Royal Society of England

Fig. 4.1 Robert Mallet (1810–1881) was an Irish geophysicist and civil engineer. He was one of the founders of seismology and was the first to use this term to denote a separate scientific discipline. He created a global catalogue of earthquakes. His analysis of the Great Neapolitan Earthquake of 1857 was significant for seismology. © Ladislav Csurma, 2023. All rights reserved

and the government. This was important because the Kingdom of Naples was not only shaken by the earthquake but also by political unrest.

Mallet summarized the results of his research in *Great Neapolitan Earthquake of 1857: The First Principles of Observational Seismology*, published in 1862. This extensive scientific report, in two bulky volumes, contained, in addition to maps and graphs, monoscopic and stereoscopic photographs documenting the damage and changes caused by the earthquake (probably the first use of then new technology for such a purpose). Mallet made use of measurements by several instruments as well as the new research methods he had developed. His map of earthquake effects was also groundbreaking. For the first time, isoseismals—closed curves that separate locations with different earthquake effects—were plotted on such a map. Also using isoseismals, he determined, for example, the location of the earthquake's origin including an estimate of the depth (the focal point or hypocentre). Mallet also tried to find surface (terrain) changes. He identified landslides over large areas, rock falls and lateral displacements in rock layers.

Mallet compiled a global catalogue of earthquakes in 1854 based on collected direct observations of earthquake effects. The catalogue and the corresponding map served as the most comprehensive representation of seismic activity worldwide for almost half a century. In 1858, Mallet was the first to introduce the term "seismology" to refer to the independent science dedicated to the study of earthquakes using the latest knowledge from mathematics, physics, and geology. He also introduced other important concepts that are still used in seismology today, including the focus of an earthquake—the point where the earthquake originates. He is rightfully considered one of the founders of seismology.

It is worth noting that isoseismals, in the 1880s, allowed indicating that earthquakes in the Eastern Mediterranean region had unusually deep hypocentres. This is because if the intensity decreases very slowly with increasing distance from the location of its maximum value, it indicates a great depth of the earthquake's focus. Isoseismals are still used today to display the geographic distribution of earthquake effects.

Mallet's method to numerically evaluate the intensity of the effects of earthquakes, represented by isoseismal maps, was replaced after 1880 by the Rossi-Forel 10-degree intensity scale and later by the Mercalli scale, which became the predecessors of today's intensity scales.

A Breakthrough Idea

Although observations of changes on faults in connection with earthquakes gradually accumulated, they were mostly understood as consequences of earthquakes rather than their causes. It was not until the end of the nineteenth century that a different view on what happened emerged. In 1891, the Mino-Owari earthquake in Japan, which resulted in thousands of casualties, left a scar on the Earth's surface in the form of a more than 100 km long rupture. Professor Bunjiro Koto (1856–1935) thoroughly examined the rupture and concluded that the rupture was the cause, not the result, of the earthquake. This was a groundbreaking idea but was only widely accepted after Reid's theory, which we discuss in the chapter on San Francisco.

Richard Dixon Oldham

Oldham, whom we mentioned back in the second chapter, published an extensive study in 1899 about one of the most devastating earthquakes that had occurred in the Assam province of India in 1897. According to current estimates, it had a magnitude of Mw 8.1–8.3. Oldham documented significant terrain deformations and demonstrated movements along faults. He found the largest uplift to be over 10 m. The vertical accelerations of seismic motion were so significant (greater than 1 g; recall that "g" represents the gravitational acceleration on Earth's surface, which is approximately 9.81 m/s^2) that they propelled boulders into the air, leaving holes in the ground with intact side walls. The intense acceleration threw many people to the ground.

In his more than 500-page study, Oldham thoroughly documented, described, and analysed available observations. Given the progress in seismograph development during this period, he was able to include the first instrumental earthquake records in his work, specifically from Italian seismic stations. His book became a valuable source of information for his successors. Even recent studies performed in 2001 and 2021 revisiting this earthquake with the modern scientific background, built upon Oldham's work.

Development of Instruments for Recording Earthquakes

No matter how complete and good, the analysis of visible earthquake effects could not by itself lead to a sufficient understanding of what is happening inside the Earth, how earthquakes occur, and how shaking occurs and behaves on the Earth's surface. A condition for understanding and for the emergence of a modern scientific discipline based on objective measurements was instruments that could accurately record nearby and distant, small and large earthquakes. The journey from Chinese seismoscopes to current seismometers with a sensitivity that we cannot "sensorily" imagine was very long and intricate.

Chinese Seismoscope

The seismoscope was invented by the Chinese scholar Zhang Heng (78–139), who lived during the Han Dynasty (202 BC–9 AD, 25–220 AD). Zhang Heng was exceptionally versatile and made contributions to various fields of knowledge and technology. An esteemed astronomer, mathematician, philosopher, inventor, and government official, he is often referred to as the Leonardo da Vinci of ancient China.

The seismoscope was one of his most significant inventions, and he introduced it at the imperial court in the capital city of Luoyang in 132 AD. While

Fig. 4.2 A seismoscope constructed by Chinese mathematician, astronomer and geographer Zhang Heng. In a sudden movement due to an earthquake, a small ball fell out of the dragon's mouth and its impact into the frog's mouth produced a sound signal. The direction of the first movement could be approximately indicated by the dragon's head from which the ball fell out. Although it does not appear at first sight, fine-tuning the attachment of the balls in the mouths was probably very difficult. © Ladislav Csurma, 2023. All rights reserved

there are no surviving physical remains or contemporary illustrations, historical documents describe the seismoscope as a large bronze device, shaped like an urn or a vase (Fig. 4.2). On its outer surface, there were eight dragon heads with bronze balls held in their jaws. Beneath each dragon, there was a frog with an open mouth, meant to catch the falling ball. Each dragon and frog were oriented according to the cardinal directions, so that the government would know where to send aid in the event of an earthquake. Initially, Heng's invention was met with scepticism, in part because he wasn't popular among his colleagues.

However, a few years later, one of the balls fell. In the capital city where the seismoscope was located, no one felt anything. But when a messenger arrived with news of a massive earthquake hundreds of kilometres west of Luoyang, precisely in the direction of the empty dragon's jaw, opinions about the seismoscope immediately changed.

From today's perspective, Heng's seismoscope was a technological marvel of its time. The technical solution of the seismoscope commands respect even today. Undoubtedly, Heng's seismoscope greatly encouraged other scholars in their efforts to develop an instrument to record earthquakes. However, humanity had to wait almost nineteen centuries for seismographs, thanks to which seismology could begin to develop based on objective measurements.

Seismoscopes in Europe and America

Mentions of earthquake detection devices in Europe are known only since the beginning of the eighteenth century. In 1703, the French Catholic priest, physicist, and inventor Jean de Hautefeuille (1647–1724) proposed filling a bowl to the brim with mercury. During an earthquake, the mercury was supposed to spill in one of eight main directions, be captured in the corresponding receptacle, and thereby determine the direction of the tremor.

An important step was the attempt to utilize the inertia of a pendulum so that it would move relatively to the Earth during an earthquake. Italian Nicholas Cirillo was perhaps one of the first Europeans to use a simple pendulum in studying a series of earthquakes in Naples in 1731.

Periods of increased efforts in constructing earthquake detection devices in the eighteenth century were often responses to specific earthquakes. In affected areas, common people made such efforts for their own protection as well. For example, in Italy, the strong interest in earthquake detection followed the occurrence of six Calabrian earthquakes in 1783, which resulted in massive loss of life and property. Italian writer and politician Francesco Saverio Salfi (1759–1832) wrote that ordinary residents used bowls filled with liquid and finely balanced objects as simple seismoscopes.

In the years immediately following the earthquakes, at least two seismic instruments were described. The Italian cleric and astronomer Atanasio Cavalli (1729–1797) rediscovered Jean de Hautefeuille's seismoscope with a mercury-filled bowl in 1784, probably without knowing the original description. The Italian naturalist, physicist and volcanologist Ascanio Filomarino, third Duke of Torre (1751–1799) added a timing device to the ordinary pendulum. He obtained several records consisting of one or two lines written in pencil for each earthquake. However, they contained just little more information than could be obtained from 'natural seismograms', such as the traces left by heavy furniture moving over a smooth surface. It is not clear whether the timing device actually worked.

In response to earthquakes in the New Madrid region, USA, at the turn of 1811 and 1812, Jared Brooks of Louisville, Kentucky, constructed pendulums of various lengths (from 2.54 cm to 15.24 cm) to record earthquakes. This was because different lengths were sensitive to movements at different frequencies. Daniel Drake, a physician from Cincinnati, Ohio, described and catalogued the sequence of earthquakes in detail, using a homemade pendulum seismoscope to determine the direction of movement.

In 1839, a series of small earthquakes began over several years near the town of Comrie in Perthshire, Scotland. As a direct result, a special committee of the British Association for the Advancement of Science was set up to obtain 'instruments and registers for the registration of earthquakes in Great Britain'. An important experiment was the inverted pendulum device designed by the Scottish physicist and glaciologist James David Forbes (1809–1868).

In all cases, the commitment to progress in earthquake measurement waned significantly relatively soon after the local earthquake period.

From Seismoscopes to Seismographs

Several significant improvements emerged in the years 1850 to 1870. In order to study the frequency content of seismic waves, Italian physicist P. G. M. Cavalleri constructed six short pendulums with different periods in 1858, each of which left a record of its motion in fine powder. He probably didn't know that pendulums of various lengths had already been constructed 40 years prior by Jared Brooks for observing earthquakes in the New Madrid area.

A significant advancement was made by Italian physicist Luigi Palmieri (1807–1896) when he proposed a mercury-based device with earthquake time recording in 1856. This device was evidently quite sensitive and was also later used to trigger the recording devices of more modern seismographs.

A technical innovation came with the design of a horizontal pendulum in 1869 by German astrophysicist Johann Karl Friedrich Zöllner (1834–1882). The instrument was installed in the basement of the university in Leipzig, and Zöllner recorded significant pendulum movement, even as a result of the filling of the lecture hall on the second floor of the building. Although the pendulum was originally constructed to observe changes in the direction of gravity due to tidal effects, Zöllner proposed its use as a seismograph.

The credit for designing the first seismograph in the modern sense of the word (i.e., a device that accurately records the time course of seismic motion) is attributed to Italian physicist Filippo Cecchi (1822–1887). His device was put into operation in 1875 but was so insensitive that it recorded its first earthquake only 12 years later. Consequently, the pioneering efforts eventually belong to a group of British professors at the Imperial College of Engineering in Tokyo, Japan, at the end of the nineteenth century. The leading figure among them was English geologist and mining engineer John Milne (1850–1913).

After the great Yokohama earthquake, the Japan Seismological Society was founded in the spring of 1880 at Milne's instigation. This was the first society devoted to seismology. At the time of its foundation, the Society had more foreign than domestic members. The instrumental innovations of the 1880s were almost entirely the work of visiting British professors. Sir James Alfred Ewing (1855–1935) and Thomas Lomar Gray (1850–1908) designed improved instruments to replace the Palmieri seismographs previously used in Tokyo. These were spring-loaded vertical seismographs and horizontal pendulum seismographs, which were set up to record displacements during earthquakes.

On 3 November 1880, these instruments recorded a small local earthquake, resulting in the first sufficiently long records of seismic motion over time. In his report on this and four other small earthquakes recorded in the same month, Ewing described the first seismograms as follows: "(1) The very gradual beginning and ending of the disturbance. In none of the observations did the maximum motion occur until after several complete oscillations had taken place. (2) The irregularity of the motion. The successive undulations are widely different both in extent and in periodic time. (3) The large number of undulations in a single earthquake, and the continuous character of the shock. (4) The extreme minuteness of the motion at the Earth's surface."

For the first time, scientists had access to a visualisation of the earthquake motion, which revealed a different mode of vibration than previously thought. In particular, it was shown that Robert Mallet's widely accepted idea that earthquakes consisted mainly of longitudinal impulses was incorrect. Only instruments capable of recording the transverse component of the motion made this discovery possible.

For the first time in history, seismologists could begin to design their instruments with more purpose and with some knowledge of the phenomena they were intended to record.

Seismograph designers were faced with three fundamental problems. The first was sensitivity, the second was friction during recording, and the third was damping the oscillations of the seismometer's inertial mass.

It was already clear that earthquake research needed instruments that could record not only those earthquakes that could be felt by humans, but also those that could not. In other words, small earthquakes nearby and large earthquakes far away. Getting the instruments to be sensitive enough has been a challenge. A good example is the SEIS seismometer on Mars, about which we write in the last chapter. Its sensitivity is amazing.

Similarly, it is understandable that a time record only makes sense if the displacement at a given time corresponds solely to the motion of the site at that time. Let's imagine a frame fixed to the surface of the Earth. A ball is suspended from the frame by a thin string. When the surface of the Earth moves, the ball tries to stay in place, following the principle of inertia. The movement of the frame relative to the ball is obviously a consequence of the movement of the Earth's surface. If we record this, we have a record of the movement of the Earth's surface. The problem is obvious: after the first short movement of the Earth's surface, the ball would start to swing for a long time. If the Earth's surface continued to move, the relative motion of the frame to the ball would no longer correspond solely to the motion of the Earth's surface. So, the oscillation of the ball has to be damped so that the instantaneous motion of the ball corresponds only to the instantaneous motion of the Earth's surface. It took a long time to devise clever damping, and it was not until the beginning of the twentieth century—a time when the theory of relativity and the foundations of quantum physics were emerging—that it was achieved.

In 1902, the German physicist and seismologist, officially the world's first Professor of Geophysics, Emil Wiechert (1861–1928) constructed an inverted pendulum instrument that was the first to incorporate proper damping. The eminent Russian physicist Prince Boris Borisovich Golitsyn (1862–1916) constructed a second pioneering type of instrument. He used the principle of electromagnetic induction with photographic recording, which made it possible to construct compact instruments with high sensitivity and precision. Photographic recording eliminated the problem of friction. Golitsyn's instrument was initially unable to compete with mechanical seismographs, but after some improvements it became the dominant type of seismograph and remained so until about 1970.

The next significant technological leap was associated with the advent of computers and digital recording.

To conclude this section, let's mention one of the oldest seismic stations in Europe, the seismic station in Hurbanovo in Slovakia (historical names: Ógyalla, Stará Ďala). From 1 January 1902, it was equipped with a pair of Strasbourg horizontal pendulums by J. & A. Bosch. In 1903, the Vincentini-Konkoly pendulum was installed. Between 1909 and 1912, Bosch pendulums were replaced with Mainka seismographs. The last seismic event recorded by the Vincentini-Konkoly seismograph dates back to 1911. Due to the disturbances of World War I, records from 1913 to 1918 were lost. The Mainka seismographs were reconstructed in 1927, and new dampers were installed in 1936. The original mechanical seismographs in Hurbanovo continue to record earthquakes to this day. The seismic station in Hurbanovo is a unique facility in Europe and globally, as functional instruments of this type are very rare in the world.

Linking with Physics Theories of Wave Propagation

The foundations of theoretical research in seismology were based on physical studies in the field of the theory of the elasticity of solid bodies, many of which were motivated by the study of the propagation of light. At the time, light was thought to be a wave in an invisible elastic ether. In the nineteenth century, the work of the French mathematicians Augustin-Louis Cauchy (1789–1857) and Siméon Denis Poisson (1781–1840) was particularly important for the emerging field of seismology. Around 1830, Poisson discovered that two types of waves—longitudinal and shear—should propagate at different speeds in solid bodies. He also estimated the ratio of their propagation speeds. Other scientists continued to study wave phenomena. For example, the English mathematical physicist George Green (1793–1841) studied wave transitions at boundaries, and the Irish physicist and mathematician Sir George Gabriel Stokes (1819–1903) studied the radiation of waves from a spatially confined source.

The theoretical results of Cauchy, Poisson, Green and Stokes were attempted to be applied to seismic research in the 1850s by the English mathematician and geologist William Hopkins (1793–1866) and the aforementioned Robert Mallet. Hopkins proposed the theoretical principle that the arrival times of the first waves could be used to locate the epicentre of an earthquake. Mallet went further; he wanted to study earthquakes using seismic waves generated by the earthquakes themselves. He emphasised the need for precise quantitative observations of the directions and magnitudes of seismic motions. He also carried out experiments in which he used explosives to generate seismic waves in order to determine the speed of seismic waves in the Earth. His experiments foreshadowed modern exploration seismology. However, the low sensitivity of the instruments at the time prevented him from obtaining accurate speeds.

The first seismographs and the records they produced naturally led to efforts and the need to interpret the records in terms of the longitudinal and shear waves theoretically discovered by Poisson and the surface wave theoretically found by the British mathematician and physicist Lord Rayleigh (John William Strutt, 1842–1919, Nobel Prize for Physics in 1904).

By the end of the nineteenth century, there was a consensus that a significant part of the Earth's interior was solid and that all three types of waves should propagate within it. Since seismologists of the time already assumed, and partially knew, that the Earth's interior was heterogeneous, it was clear that interpreting earthquake records would not be straightforward. They cautiously distinguished between the "preliminary tremor" and the "main phase". They estimated that these consisted of a group of longitudinal and shear waves and a group of surface waves, respectively. In 1899, Oldham identified two phases corresponding to longitudinal and shear waves within the first group in the records of the Assam earthquake in India. He identified the motion within the main phase of the earthquake record as a surface wave.

Global Seismology

From a certain perspective, the "birth" of global seismology can be traced back to the first seismic recording of a distant earthquake, to which a fortunate coincidence also contributed. In 1889, the German astronomer and geophysicist Ernst Ludwig August von Rebeur-Paschwitz (1861–1895) attached horizontal pendulums with photographic recording to the bases of telescopes in Potsdam and Wilhelmshaven (240 km apart). His aim was to measure the movements of the Earth caused by the gravitational influence of celestial bodies such as the Moon and the Sun. The "disturbance" that occurred simultaneously at both stations found its surprising explanation when he read a report in the journal *Nature* about a major earthquake in Japan. Although Milne had hypothesised in 1883 that, with suitable equipment, any significant earthquake could be recorded anywhere on Earth, it was Ernst von Rebeur-Paschwitz's recording in Potsdam that confirmed this possibility.

The first international seismological conference was held in 1901 in Strasbourg (then Germany, now France). In 1904, the International Seismological Association (ISA) was founded with a central committee in Strasbourg to promote research and international cooperation in seismology. In 1922, the ISA became the Seismological Section of the International Union of Geodesy and Geophysics (IUGG), founded in 1919. This section later became the IASPEI (International Association of Seismology and Physics of the Earth's Interior), which is still part of the IUGG. In 1949 some European scientists, led by Inge Lehmann, prepared a project which later merged to the formal creation of the European Seismological Commission in 1952. In 1964, the International Seismological Centre (ISC) was established in the United Kingdom to collect earthquake data from around the world.

Research of earthquakes and the study of the Earth's internal structure is not possible without global cooperation and coordination. There is also a mathematical and physical reason for this: in order to model the Earth's interior and its processes quantitatively, it is necessary to have the densest possible coverage of the Earth's surface with stations.

5

San Francisco 1906: USA vs. Earthquakes

The Great San Francisco Earthquake, which nearly destroyed San Francisco in 1906, is one of the most significant earthquakes in recorded history. While it remains one of the most destructive earthquakes in the United States, its significance lies primarily in the way it changed the understanding of earthquakes and the approach to seismic hazard. It can be said that the San Francisco earthquake is comparable to the Great Lisbon Earthquake and Tsunami of 1755 in terms of progress in understanding what happens inside our planet.

An Earthquake That Almost Destroyed San Francisco

The Great California Earthquake occurred on 18 April 1906 at 5:12 AM on the San Andreas Fault, just a few kilometres from the Golden Gate Bridge in San Francisco. The earthquake ruptured approximately 480 km of the San Andreas Fault, stretching from the town of San Juan Bautista (south of San Francisco) to Cape Mendocino (north of San Francisco). Current estimates put the earthquake's moment magnitude (Mw) at around 7.9. The effects of the earthquake were felt in an area roughly 1200 km by 550 km, roughly bounded by the city of Los Angeles in the south, the town of Winnemucca in Nevada in the east and the southern parts of the state of Oregon in the north. Fortunately, the area of the greatest damage was much smaller, consisting of a strip a few tens of kilometres wide along the ruptured part of the fault.

The city of San Francisco was one of the hardest hit areas. Many buildings were damaged by the earthquake. Shockingly, shortly after the earthquake, a

P. Moczo et al., *Earthquakes*, Springer Praxis Books, https://doi.org/10.1007/978-3-031-64707-9_5

Fig. 5.1 Aerial view of San Francisco destroyed by the earthquake and fire (28 May 1906, 40 days after the earthquake). USGS—public domain

fire began to spread through the city and its ruins, and continued to ravage the city for another 3–4 days (Fig. 5.1). The fire was caused by ruptured gas mains and exacerbated by the use of dynamite to demolish buildings damaged by the earthquake. In total, some 28,000 buildings were destroyed, leaving up to 225,000 of the city's 400,000 inhabitants homeless. It is estimated that the total economic damage caused by the earthquake and fire exceeded 400 million US dollars (equivalent to more than 13 billion US dollars today), with the earthquake accounting for only 10–20% of the direct damage. Initial reports (including testimonies) and estimates indicated a relatively low number of casualties (around 700 victims). However, subsequent analysis of archival records suggests that the death toll may have been as high as 3000.

It is worth noting that politicians and businessmen used the scale of the fire to argue that it was the fire, not the earthquake, that destroyed the city. The media misinterpreted the views of scientists. Behind the scenes, politicians and businessmen were worried about scaring off potential investors by earthquake risk.

However, the shocking experience of the fire led to a significant improvement in fire safety in the city. A long-established and well-maintained fire protection system proved effective during the 17 October 1989 earthquake with its hypocentre near Loma Prieta, the highest peak in the Santa Cruz Mountains. Even during this earthquake, several fires broke out almost immediately, but were quickly extinguished.

Although the city's fire protection was improved, the 1906 earthquake and its analysis unfortunately did not lead to any meaningful planning for earthquake preparedness. This did not happen until after the 1989 Loma Prieta earthquake, when the Seismic Hazard Mapping Act was passed, requiring the publication of maps showing areas of expected seismic ground motion amplification, soil liquefaction and landslide potential. An Earthquake Resilience Programme was initiated, with over 30 billion US dollars allocated for earthquake preparedness.

An Exceptional Interest of Scientists

Earthquakes were relatively common in California. Between 1836 and 1906, up to 16 earthquakes with a magnitude of more than 6 occurred in the San Andreas Fault region around San Francisco. As a result, prominent scientists and other residents of the area kept relatively detailed observation diaries. Earthquakes were essentially considered part of everyday life. Andrew C. Lawson (1861–1952), professor of geology at the University of California, Berkeley, stated in 1904 that Californian earthquakes were annoying and inconvenient, but based on historical experience, they were not dangerous.

The great earthquake of 1906 and its aftermath were all the more surprising and aroused great interest in the scientific community. Many scientists, not only in California, began to study the earthquake. Lawson quickly recognised the need to coordinate earthquake research. On his initiative, Governor George C. Pardee appointed the State Earthquake Investigation Commission just three days after the earthquake. Lawson served as chairman. Commission members included geologist Grove Karl Gilbert (1843–1918) of the United States Geological Survey (USGS), geophysicist and seismologist Harry Fielding Reid of Johns Hopkins University, geologist John Casper Branner (1850–1922) of Stanford University, and astronomer Armin Otto Leuschner (1868–1953) of the University of California, Berkeley. The Commission coordinated the work of more than 25 geologists, seismologists, geodesists, engineers and more than 300 other people who participated in the Commission's activities.

The Commission published a preliminary report less than 2 months after the earthquake, on 31 May 1906. The main findings were later published in two volumes. The first, coordinated by Lawson, was published in 1908 and was mainly devoted to documenting the findings—from geological discoveries to eyewitness accounts. It includes many photographs and maps. The

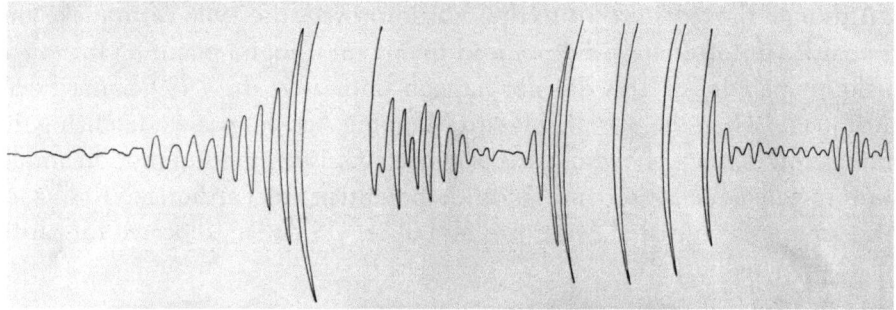

Fig. 5.2   Redrawn recording of the north-south component of the San Francisco earth-quake motion at the Ógyalla (now Hurbanovo, Slovakia) seismic station in an article published by Jenő Cholnoky in the Bulletin de la Société Hongroise de Géographie in 1906. This is a recording from a horizontal seismometer by J. & A. Bosch. It can be seen that the dynamic range of the motion exceeded the possibilities of graphic recording. Reprinted with permission from Moczo et al. (2023), © GRADA Slovakia s.r.o., 2023. All rights reserved

second volume, coordinated by Reid, was published in 1910 and focused on the seismological and mechanical aspects of the earthquake.

The report presents a hypothesis about the origin of the earthquake and provides a unique systematic and comprehensive summary of observations and information about the earthquake. It also contains detailed information on seismometers at nearly 100 seismic stations around the world (Fig. 5.2), including the arrival times and amplitudes of the seismic waves recorded at each station. It also includes copies of seismograms from 72 stations.

The Construction Quality

The coordination of damage documentation after the earthquake and the concentration of different types of observations in one place led to the recognition of important peculiarities.

Considering the large area affected by the earthquake, the damage to buildings was surprisingly relatively concentrated. Even within San Francisco itself, it was possible to identify areas that were significantly more affected by the earthquake than others. For example, there was significantly more damage in the harbour area—in places where landfill had been used to fill in the marshland.

A common characteristic of the sites severely affected by the earthquake was that they were located on the surface of sediment-filled valleys or basins or on a landfill. Partially or completely confined sediments not only amplified but also prolonged the shaking.

Another important observation was that the quality of construction had an impact on the damage caused by the earthquake. Many well-built wooden structures survived the earthquake without significant damage but were destroyed by the subsequent fires. In San Francisco, the first high-rise steel buildings were constructed, such as the 19-storey Klaus Spreckels Building and the 16-storey San Francisco Chronicle Building. Both buildings survived the earthquake with minimal damage but were damaged in the fires that followed. In the harbour area, where the earthquake caused the most damage in San Francisco, buildings built on piers were virtually undamaged. Similarly, the tracks of the cable car system were also laid on deep pilings, which is why they remained passable even in areas affected by soil liquefaction. In many cases, they were the only means of transport as roads and sidewalks were destroyed.

San Andreas Fault (and a Network of Faults in California)

Although the San Andreas Fault is now the most famous earthquake fault in the world, before the 1906 earthquake only a small part of it had been mapped in the San Andreas Valley near San Francisco. This section had been mapped by Lawson in 1895. Moreover, at that time it was only a matter of mapping the geological structure, with no reference to the origin of earthquakes.

The distribution of surface cracks and ruptures, and the displacement of fences, roads, and other linear structures after the 1906 earthquake led to the realisation that it was part of a larger structure. Shortly after the earthquake, the San Andreas Fault was mapped over a length of about 220 km on land (although a significant part of the fault lies under the Pacific Ocean). The fault was then mapped further south, down to relatively fresh surface cracks from the 1857 earthquake near the town of San Bernardino. In total, almost 1000 km of the fault has been identified.

Today, we know that the San Andreas Fault is the primary boundary between the Pacific and North American tectonic plates. Along this fault, western California tends to move northwestward relative to the eastern part of California, at a rate of approximately 5 cm/year. The San Andreas Fault stretches for approximately 1300 km. It starts in the north at Cape Mendocino, at a junction of three plates—the Pacific, North American, and Gorda plates.

In the south, it terminates in the Salton Sea area of southern California, where the transform fault gradually transitions into a divergent boundary.

It is estimated that the fault is about 15–20 million years old, and over that time, there has been a total displacement of up to approximately 550 km along its length. There are identified locations where the displacement across the fault has reached up to 240 km.

If the San Andreas fault was a perfect straight line, and if the motion of the Pacific and North American plates was exactly in the direction along the fault, then the contact of the plates would really only be the San Andreas fault. The relative motion of the plates would only take place here, i.e., earthquakes would then only occur on the San Andreas fault. However, plates generally have irregular boundaries, and this is also true of the San Andreas fault, which is slightly "wavy". Recall that the Pacific plate is moving in a northwesterly direction, relative to the North American plate. North of Los Angeles, the San Andreas fault turns slightly to the east. Therefore, in the vicinity of Los Angeles, the relative motion of the plates is more complex and there is more deformation within the plates as well. Consequently, secondary faults have emerged (Fig. 5.3).

In the San Francisco area, the situation is seemingly a little simpler as the San Andreas fault is more rectilinear than in the Los Angeles area. However, even here the San Andreas is not perfectly straight, and therefore deformation and faulting also occurs within the plates.

Earthquakes Before 1906

Probably the most significant contribution of the San Francisco earthquake was the development of the elastic rebound theory. To appreciate this, let's first remind ourselves how scientists viewed earthquakes before 1906.

Earthquakes were considered a natural phenomenon, but the mechanism of their origin was unknown. Surface cracks, faults, and fault lines were observed in many places around the world after strong earthquakes. Although it was generally assumed that they were caused by earthquakes (i.e., unrelated to the cause of earthquakes), we now know that as early as the Great Japanese Earthquake of 1891, Bunjiro Koto realized that slip along a fault could be the cause of an earthquake. The origin of earthquakes themselves was more commonly associated with volcanic activity or the force of gravity, which was the strongest known force acting on the Earth's surface. The theory of lithospheric plate tectonics did not yet exist.

Fig. 5.3 Identified faults in California. The San Andreas Fault is marked in red. Modified from USGS—public domain

The Elastic Rebound Theory

The development of the elastic rebound theory, explaining the preparation and occurrence of earthquakes, was also facilitated by favourable or even fortunate circumstances. Prior to the 1906 earthquake, relatively extensive geodetic measurements had been conducted in the California region. Since 1851,

The United States Coast and Geodetic Survey (USCGS) had been carrying out extensive triangulation (precise measurements of the relative positions and elevations of points) also in the area affected by the 1906 earthquake. These measurements were concluded in 1892. The resulting grid was intended to serve as a highly precise reference grid for various planned measurements, such as the accurate mapping of the coastline.

After the earthquake, surface cracks with significant offsets were observed in several places. In some locations, these surface cracks severed or ruptured fences, roads, or pipelines. The largest offset reached up to 6.5 m, with an average offset of about 3 m.

Such large relative displacements could cause significant changes in the relative positions of precisely measured points, thereby reducing the overall accuracy of the reference grid. Therefore, the USCGS conducted control measurements to verify the accuracy of the grid and to determine the areas/points that would need to be measured again. Surprisingly, these measurements showed that changes in positions occurred not only at points near the fault, but also at distances of several kilometres from the fault. Based on these findings, a new triangulation was performed in an area of approximately 270 km × 80 km. Since the original measurements were made in two stages, in 1851–1865 and 1874–1892, measurements from three periods were available in partially overlapping areas. Analysis of the measurements revealed that changes in positions of the points had already occurred after the 1868 earthquake near Hayward (M6.3–6.7).

Geodesists observed and described these changes in the first volume of the aforementioned investigative commission report. However, Harry Fielding Reid attempted to provide a physical explanation in the second volume. Based on geodetic observations, he knew that the most significant changes occurred in the vicinity of the fault during the earthquake. The magnitude of the displacement decreased with distance from the fault, but changes were still noticeable at a distance of about 10 km from the fault. He assumed that the displacements must have been caused by sufficiently large forces. Since the displacements were almost purely horizontal, they could not have been caused by gravity. From observations, he knew that the displacements occurred suddenly during the earthquake itself. However, sufficiently large forces could not have arisen suddenly out of nowhere. With these considerations, he concluded that horizontal forces caused deformation and an increase in stress around the fault. During the earthquake, this deformation was suddenly released, and the material reached a stress-free state. Although at the time there was no known process that could generate sufficiently large horizontal forces, the elastic rebound theory remains, to this day, a fundamentally valid description of the mechanics of earthquake preparation and occurrence. A brief explanation of the current understanding of earthquake generation was provided in the chapter on Earthquakes – a Short Introduction for Almost Everybody.

Reid eventually published his famous article *The elastic-rebound theory of earthquakes* in the University of California bulletin in 1911.

The Establishment of the Seismological Society of America

In response to the earthquake, the Seismological Society of America (SSA) was founded on 30 August 1906. Its goal, similar to that of the Seismological Society of Japan, founded in 1880 after the Yokohama earthquake, was to shape public opinion about earthquakes, provide advice, and provide resources for earthquake research. The official minutes of the first regular meeting of the society on 20 November 1906, state that the Society was established in response to the earthquake to encourage "the acquisition and diffusion of knowledge concerning earthquakes and allied phenomena, and to enlist the support of the people and the government in the attainment of these ends."

Today, SSA is a significant international scientific society that supports the development of seismology and its applications in an effort to mitigate the danger of earthquakes and study the structure of the Earth.

The SSA publishes the most prestigious scientific seismology journals, namely the Bulletin of the Seismological Society of America (BSSA) and Seismological Research Letters (SRL). In 2021, they were joined by The Seismic Record (TSR).

After the Great California earthquake, earthquakes in the USA ceased to be merely 'annoying and troublesome'. They became the subject of serious research with the aim of minimizing damage and the number of casualties in future earthquakes.

A reference to the 1906 earthquake is also part of the SSA's official logo. The logo includes a portion of a seismic record of the earthquake recorded at the seismic station in Göttingen, Germany.

6

Chile 1960: When the Earth Rings Like a Bell

Wild West of South America

The headline doesn't exaggerate. Along more than 6000 km of the western edge of the South American continent, the Nazca lithospheric plate is subducting beneath the South American plate faster than the inhabitants of Chile, Peru, and Ecuador would wish. In the region of Ecuador, at a speed of approximately 6 cm/year, in Peru at 6 to 7 cm/year, and beneath the entire Chile at 7–8 cm/year. That is truly fast and the consequences of subduction are really wild.

The subduction zone of the Nazca Plate is one of the most important subduction zones. The geometrical structure of this zone is very complex. From southern Ecuador to central Peru and central Chile, the Nazca subducts at an angle of about 10° or less. Under the northern part of Ecuador, under the southern part of Peru, the northern and southern part of Chile at an angle of about 25–30°. Under the southern part of Peru and the northern part of Chile, it plunges to depths of more than 500 km.

The subducting of the Nazca Plate under the South American Plate not only causes the gradual uplift of the Andes, but also, and more importantly, an unusual amount of earthquake and volcanic activity along the western part of South America. The seismic activity is the result of friction between the sinking plate and the resisting plate.

P. Moczo et al., *Earthquakes*, Springer Praxis Books,
https://doi.org/10.1007/978-3-031-64707-9_6

Country of the Largest Earthquakes

All of the world's largest earthquakes have occurred in subduction zones. Chile is an exceptional country in this regard: since 1900, there have been 11 earthquakes with a moment magnitude (Mw) greater than 8 beneath its territory, at depths of up to 70 km (Fig. 6.1). Out of the 16 earthquakes with Mw greater than 8.5 worldwide since 1900, 3 occurred in Chile. An earthquake with Mw 8.5 releases approximately 2000 times more energy in the form of seismic waves than the 6 April 2009 tragic earthquake near L'Aquila in central Italy which killed more than 300 people and left more than 66,000 people homeless.

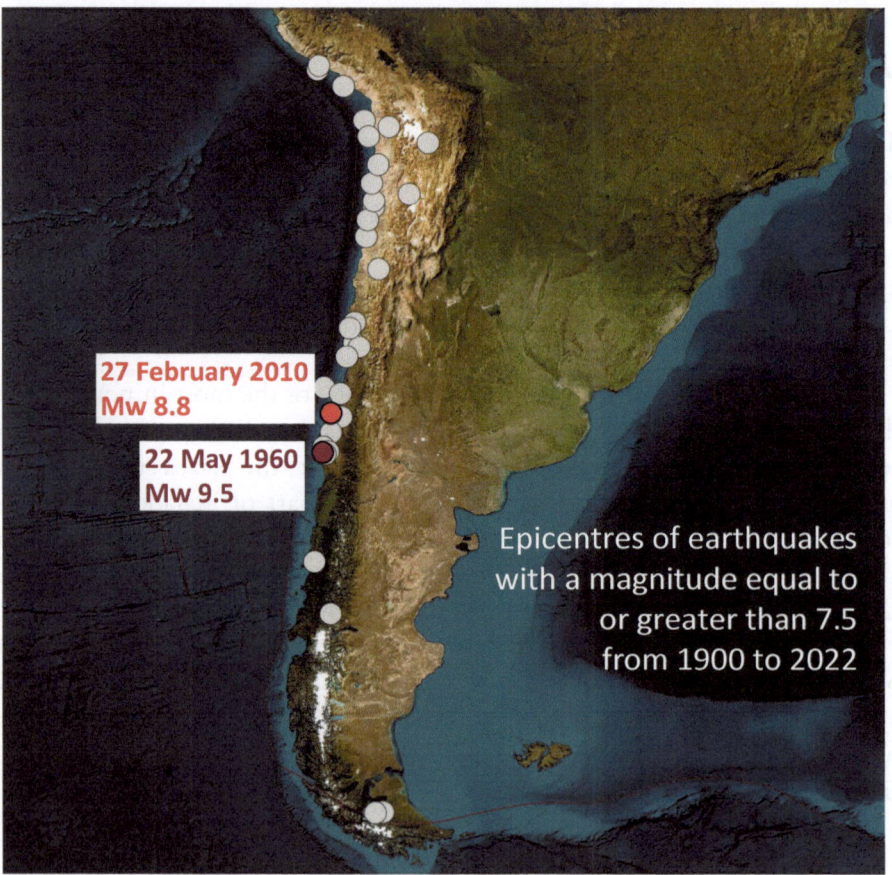

27 February 2010 Mw 8.8

22 May 1960 Mw 9.5

Epicentres of earthquakes with a magnitude equal to or greater than 7.5 from 1900 to 2022

Fig. 6.1 Map of earthquake epicentres in Chile and its immediate surroundings for the period 1900–2022. All 33 earthquakes had a magnitude greater than 7.5. Based on USGS public domain

If the epicentre of an earthquake is under the ocean, subduction earthquakes can cause tsunamis. We will explain this in the chapter on Apocalyptic Tsunamis in Sumatra and Japan.

105 Volcanoes

A total of 105 volcanoes have been identified on the territory of Chile, of which 90 are active. That's really enough for one country. Simply put, the ocean coast represents the western border of Chile, while the line of volcanoes represents the eastern border. The highest volcano, Ojos del Salado, is 6891 m high and is estimated to have erupted sometime between years 400 and 1000. The second highest volcano, Llullaillaco, is 6739 m high and last erupted in 1877. There have been 28 eruptions in Chile since 1900, with 4 volcanoes active in the single year of 2015.

The Largest Possible? The Great Chilean Earthquake

At 11 min and 20 s past three in the afternoon of 22 May 1960, the largest earthquake known to date began at a depth of 25–33 km near the west coast of Chile, some 615 km south-southwest of the capital, Santiago. It had a magnitude of 9.5. No other known earthquake has reached this magnitude. It is rightly called the Great Chilean Earthquake. It is also known as the Valdivia earthquake, named after the region where the epicentre was located.

From the hypocentre, the rupture extended over an area 965 km long and 160 km wide, i.e., 154,400 km². Over this area, the Nazca Plate has shifted on average by approximately 18 m relative to the South American Plate. Although this is a different process, suppose, for example, that the Earth's 20 km thick surface layer slides over an area 5 times the surface area of Belgium by 18 m relative to what lies beneath the layer. The energy required for such a shift is probably not intuitively or "sensually" imaginable to anyone.

Approximately, it can be estimated that this earthquake released 1.1×10^{19} Joules in the form of seismic waves. That's truly a lot. For example, approximately 175,000 times more than the atomic bomb dropped on the Japanese city of Hiroshima on 6 August 1945. Let's recall that the Little Boy bomb dropped from an airplane destroyed the city of Hiroshima and killed approximately 66,000 people 'in a second.'

No other short-term physical process on Earth releases such a large amount of energy at once as a major earthquake.

Apart from the Great Chile earthquake, only four earthquakes have reached a moment magnitude of 9.0 and above: 4 November 1952 near Kamchatka (9.0), 28 March 1964 near Alaska (9.2), 26 December 2004 near Sumatra (9.1) and 11 March 2011 near Tōhoku in Japan (9.0). All earthquakes were recorded by hundreds of seismic stations. Therefore, it was possible to determine the magnitude of the earthquake, i.e., the seismic moment and moment magnitude derived from it. However, let us emphasize that the determination of the earthquake size is not so easy. Recall that the scalar seismic moment is the product of the total area over which the rupture has spread, the average slip on this area, and the average value of the shear modulus over this area. Determining the values of the size of the ruptured area, the average slip and the shear modulus deep below the Earth's surface is a very challenging problem even with the current possibilities—the number and sensitivity of seismic stations and the power of computers.

Even more challenging is estimating the seismic moment of earthquakes in the distant past, for which only historical records of various kinds and reliability have been preserved. However, it can be inferred that since the beginning of the twentieth century, when the first seismic stations capable of recording even distant earthquakes were established, there have been no earthquakes larger than those five mentioned.

The question naturally arises as to whether an earthquake of Mw greater than 9.5 can ever occur. It is clear that the value of the shear modulus in the Earth is limited and depends on the rocks that make up the plates at the point of contact. The question is whether the rupture can extend over an area larger than about 150,000 km^2, and whether the plates can move relative to each other over such a large area of more than 18 m in average. These two questions cannot be answered because we simply do not know the geometry of the plate contacts in the subduction zones in sufficient detail, nor do we know the rheological properties of the subduction zone. The parameters of the Great Chilean Earthquake, compared to other large earthquakes, do not surprise us in the context of what we have already mentioned: in the area of the earthquake, the Nazca plate is underriding the South American plate faster (up to 8 cm/year) than in other parts of this South American subduction zone, and faster than in most subduction zones in the world. The geometry of the subduction zone is a bit simpler than geometry of other subduction zones. It is therefore possible that even if the magnitude of 9.5 is not the largest possible, it is close to the largest possible.

The Great Chilean earthquake was preceded by 4 foreshocks with magnitudes greater than 7. After major earthquakes, numerous aftershocks occur on the ruptured part of the fault contact. If the magnitude of the main earthquake is Mw, then magnitudes of the largest aftershocks usually do not exceed Mw-1. In this case, there were 5 aftershocks with magnitudes of 7 and above. Perhaps this size of aftershocks is consistent with the idea that the fault rupture extended to such an enormously large area. In other words, the fault rupture propagated relatively easily, leaving few areas of accumulated, unreleased deformation on the fault.

The Tsunami Was Killing Even 16,000 km from the Epicentre

As we have already mentioned, if there is uplift and/or subsidence of the seafloor during an earthquake, a tsunami is generated. Literally spectacular tsunamis occurred during the Great Chilean Earthquake. As in other cases where earthquakes trigger massive tsunamis, it is the tsunami itself that causes the greatest damage.

The tsunami devastated a large part of the Chilean coast. The town of Puerto Saavedra was completely destroyed. The tsunami reached a height of 11.5 m there and penetrated up to 3 km inland. It is estimated that the highest recorded tsunami height on the Chilean coast was up to 25 m. Estimating the maximum tsunami height reached is not easy. Typically, in areas where the tsunami reaches heights above 20 m, neither people nor technical equipment survive.

Tsunamis can spread imperceptibly across the Pacific Ocean and reach destructive heights even at its far edges (Fig. 6.2). After 15 h, the tsunami triggered by the Great Chilean Earthquake reached Hawaiian Islands. At a distance of more than 10,600 km from the epicentre of the earthquake, the tsunami tore two-tonne boulders from the coastal walls and destroyed the port of Hilo (Fig. 6.3). Damage in Hawaii amounted to 75 million US dollars (the current equivalent is about 774 million US dollars). The tsunami also killed 61 people (!) as many curious people came ashore despite the warning. Unfortunately, they did not know vital basics about tsunamis.

In another 7–8 h, the tsunami reached Japan and the Philippines. 32 people died in the Philippines. In Japan, as many as 138 people were killed and damage amounted to 50 million US dollars (equivalent to about 516 million US dollars today). More than 16,000 km from the earthquake's epicentre! It

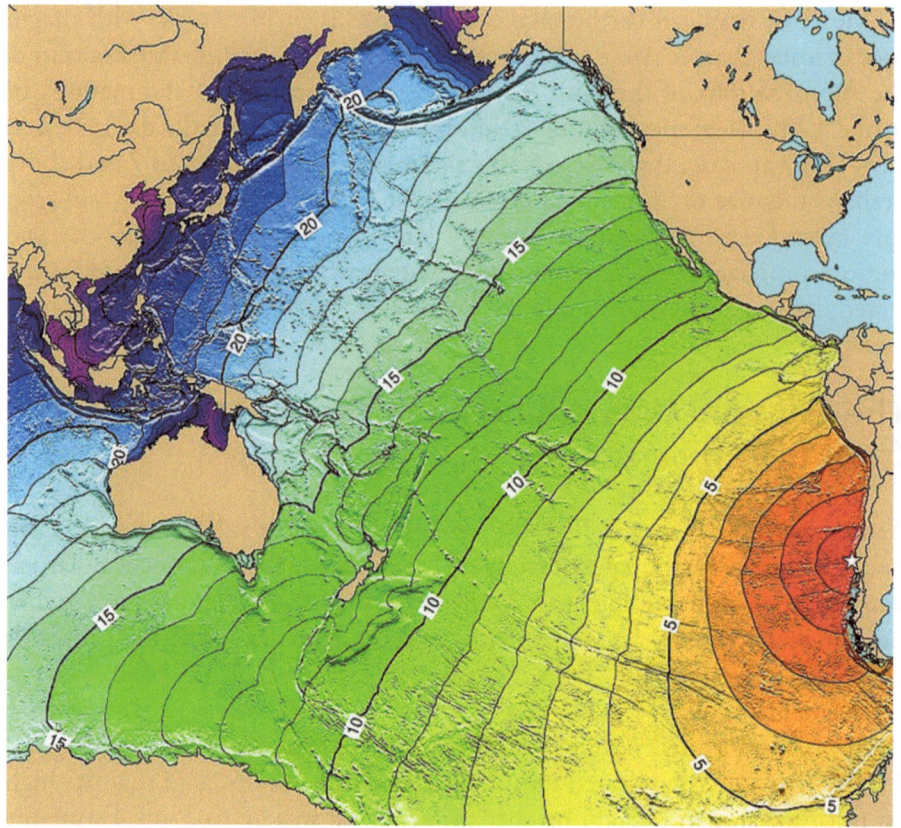

Fig. 6.2 Map of the tsunami propagation caused by the 22 May 1960 Great Chilean Earthquake. The numbers indicate the propagation time in hours from the epicentre. NOAA public domain

is highly likely that the effects of the tsunami in Japan were amplified by local conditions.

Despite its distance from the epicentre, the tsunami also caused about 500,000 US dollars damage (equivalent to about five million US dollars today) on the west coast of the USA.

Vibrating Planet

As we already know, such a large earthquake causes, in addition to intense seismic waves propagating throughout the Earth's interior, the generation of free oscillations of the Earth through constructive interference of the longest

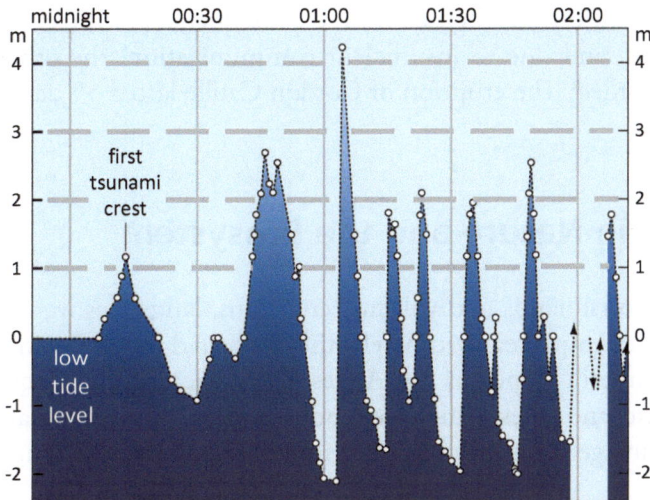

record of changes in the ocean level on 23 May 1960, Hilo, Hawaii

Fig. 6.3 Record of changes in the sea level in the port of Hilo on the island of Hawaii due to the tsunami caused by the earthquake on 22 May 1960, off the western coast of Chile. Modification from USGS public domain

waves. As mentioned earlier, the free oscillations of the Earth are analogous to the modes of vibration of a bell. Analysis of seismic records from the Great Chilean Earthquake provided seismologists with the first clear evidence that the earthquake induced free oscillations of the Earth. More than 40 modes of the oscillations were identified, with the longest period being 54 min. It is worth noting that advances in seismometers and global earthquake monitoring now allow the identification of more than 1400 modes of free oscillations of the Earth generated by the largest earthquakes.

By providing the first reliable evidence of the existence of the free oscillations of the Earth, the Great Chilean Earthquake marked the beginning of intensive observational and theoretical research into Earth's free oscillations. The methods of calculation and analysis of the oscillations laid the theoretical foundation for the emergence of helioseismology and ultimately astroseismology.

Volcanic Eruptions

Approximately 38 h after the earthquake, the Cordón Caulle volcano in the Andes in southern Chile erupted. It is the only recorded eruption associated with the earthquake. However, other volcanoes probably also erupted. Because everyone's attention was focused on the consequences of the earthquake and tsunami itself, and due to insufficient communication, the other eruptions were not recorded. The eruption of Cordón Caulle lasted 59 days and caused no loss of life.

Changes in Nature and the Ecosystem

In addition to tsunamis and volcanic eruptions, landslides were a secondary effect of the earthquake. Fortunately, they occurred especially in areas of the forested mountain slopes of the Andes above the Liquiñe-Ofqui fault, the Chilean coasts and lakes, which were only sparsely populated. Landslides, for example, damaged a section of the international road linking Chile and Argentina.

As a result of the earthquake, the landslide blocked the San Pedro River, an outflow from Lake Riñihue. The height of the resulting dam reached 24 m. Water began to accumulate in the lake at an alarming rate: every metre by which the level rose meant 20 million m³ of water not being drained. The San Pedro River had a flow capacity of 400 m³/s. Once the dam created by the landslide had been breached, 480 million m³ would have spilled into the San Pedro riverbed and its surroundings. The water would have flooded an area inhabited by 100,000 people.

It took a whole month to eliminate this threat. The army, power plant workers and public workers had to dig drains and drainage channels, which they used to gradually drain the accumulated water.

As a result of the earthquake, the average elevation in the affected area rose by 1 m, and on Guafo Island by up to 3 m. The Valdivia area experienced a subsidence of 1.5–2.2 m. The increase in the area inundated by the subsequent tsunami was due to the surface lowering.

More than 5000 ha of agricultural land and 2800 ha of urban areas remained flooded. They are still exposed to flood risk. The soil flooded by seawater was subject to salinization. The salt content in the soil increased to the extent that it became an unsuitable environment for cultivating most plant species.

On the other hand, areas permanently flooded, for example those around the Cruces River, allowed the formation of a new marshland ecosystem, subsequently inhabited by endemic South American species such as the black-necked swan or the large-flowered waterweed. Since 1981, the ecosystem has been a natural reserve called Santuario de la Naturaleza Carlos Anwandter.

The earthquake also altered river channels and the shape of the coastline, necessitating adjustments to maps.

Economic Damage in Chile

The immediate financial damage that Chile had to deal with was estimated at 550 million US dollars, which represented 12% of GDP. In 2023, this amount would be equivalent to 5.6 billion US dollars.

Almost every sector of the economy has been affected, from crop and livestock farming to energy. This was partly because, before the earthquake, the Valdivia region concentrated 50% of livestock farming and was the centre of Chile's production of condensed and powdered milk. It produced 90% of cheese products, wood and paper products, 80% of wool products and 100% of window glass and pig iron.

The area macroseismically affected by the earthquake had 48% of the total rural population, 31% of the energy production, and out of the 89 firms, 30 were of national importance. The Chilean government's efforts to rebuild the national economy were unsuccessful—not only was the damage too severe and the government too unstable, but the economic losses caused by the earthquake were not the only financial problem of Chile.

Earthquake, Tsunami, Politics, and People

In the past, known large earthquakes with a magnitude of 8 and above often caused a large number of deaths, sometimes up to hundreds of thousands. Fortunately, this was not the case with the largest documented earthquake. 1655 people died, and over 3000 were reported missing.

The total population of Chile was approximately 7.7 million in 1960. One-third of the population directly felt the earthquake. More than a quarter of the population found themselves homeless.

Chile was a country of poverty and heavy manual labour. Too many people were dependent on wages that were barely enough to provide food. The number of inhabitants was on the increase and the number of job vacancies was decreasing. At the turn of the 1950s and 1960s, a strange and confusing period occurred: young people had an increasingly high-quality and comprehensive education, but they had trouble finding employment even in positions that were significantly below the level of their formal education. Inflation was persistently sweeping the country, and the solution to the economic issue was a key point of every election program.

Unfortunate was not only the economic situation but also situation in terms of the global foreign policy. In Chile itself, the Communist Party had more popularity and stability than in other Western countries, which is why the American CIA intervened in Chilean political affairs. In early May 1960, the Cold War tensions between the United States and the Soviet Union were escalating. The Soviets shot down an American spy plane. On 18 May, the Chairman of the Council of Ministers of the Soviet Union Nikita Khrushchev (in office from 1958 to 1964) said: 'Wouldn't it be better to take the American aggressors by the scruff of the neck also and give them a little shaking?'

Even 22-year-old Chilean farmer José Argomedo learned about the relationship between the United States and the Soviet Union from the radio. When 4 days later he suddenly felt strong tremors, he thought that a nuclear war had broken out. Ultimately, he could only testify about his experience because, at the time when the tsunami flooded the coast, he happened to be in an elevated place.

Ramón Atala, then one of the most prosperous traders in the town of Maullín and owner of several businesses, escaped the dangers of the earthquake and the first crest of the tsunami. As the water receded, he gathered the courage to return to one of the warehouses to save the cash stored there. However, the second crest of the tsunami swept away the entire warehouse. Ramón Atala's body was never found.

Many people saved themselves from the tsunami by climbing a tree or the roof of a house. They could survive even if those were carried by the water. Whole houses were swept off the mainland by the tsunami and bodies of the dead were washed out of cemeteries. Many of the tsunami victims were in turn covered by sand deposits.

The earthquake truly shook the lives of all. A few days after the earthquake, the Lafkenches tribe of the Mapuche indigenous people of the coastal village of Collileufu performed a human sacrifice at the behest of a local shaman— called 'machi' in Chilean indigenous culture—Juana Namuncura Añen. A 5-year-old boy, José Luis Painecur, was sacrificed with the intention of making peace with the Earth and the ocean.

Local authorities only became aware of the sacrifice through reports of the theft of two horses—which were to be ritually consumed during the sacrifice. Two local men were convicted of the murder. They were released from custody after 2 years. The argument was that they had 'acted without free will, driven by an irresistible natural force of ancestral tradition'.

7

Incredible Mexico 1985: Destruction at a Distance

How the Capital of Mexico Was Founded

The Aztecs In the north of present-day Mexico, there was the legendary land of Aztlán, inhabited by Nahuatl-speaking peoples of Nahua ethnicity. From the ninth century, these peoples gradually left Aztlán. The last of them were the Aztecs, who settled in the Valley of Mexico after two to three centuries of migrating. Today, the Valley of Mexico encompasses a large part of the territory of the capital city, Ciudad de México, and parts of the federal states of México, Hidalgo, Tlaxcala, and Puebla.

The Aztecs did not arrive in an uninhabited territory. For hundreds of years before their arrival, the Valley of Mexico was inhabited by a civilization that disappeared in the eighth century for unknown reasons. Evidence of its existence can be found in the ruins of the city of Teotihuacán, located about 50 km from today's capital. After the decline of this civilization, the Toltecs dominated the Valley of Mexico until their decline around the twelfth century. During these centuries, large groups began to penetrate this area, bringing with them a new concept of the city-state as a territorial-political entity. This led to the emergence of relatively large city-states scattered throughout the valley. In terms of population, the most significant of these was México-Tenochtitlan, a relatively small city-state located on an island in the western part of Lake Texcoco, the largest of the five lakes in the Valley of Mexico. México-Tenochtitlan was probably founded in 1325 by the Aztecs, who began to concentrate there in the thirteenth century after several hundred years of migration.

P. Moczo et al., *Earthquakes*, Springer Praxis Books,
https://doi.org/10.1007/978-3-031-64707-9_7

The location of the newly formed homeland was allegedly chosen after a long search based on a sign from the supreme god Huitzilopochtli, the god of war, the Sun, and human sacrifice, who eventually also became the patron of the city. They were to find their new home where they saw an eagle perched on a cactus and devouring a snake. This scene is still depicted on the Mexican flag. They believed this place was where the world was reborn, and a new Sun was created after the last cosmic catastrophe.

The natives did not welcome the presence of Aztecs. The city-state of México-Tenochtitlan expanded and subjugated surrounding city-states and their native populations over the following century. In 1428, a confederation of three city-states was formed: México-Tenochtitlan, Texcoco, and Tlacopan, known as the Aztec Triple Alliance. The Triple Alliance gradually gained dominance over the other city-states, which were weakened by internal conflicts. From the Triple Alliance, México-Tenochtitlan emerged as the leading state before the arrival of Spanish conquerors in the sixteenth century.

At the beginning of the sixteenth century, during its peak, the Aztec Empire, the territory under Aztec influence, stretched across much of Mesoamerica, with the Gulf of Mexico to the east and the Pacific Ocean to the west. The population of the Aztec Empire is estimated at six million, with approximately 200,000 residing in the city of México-Tenochtitlan. In terms of population, the city ranked among the largest metropolises of the time. It had tall buildings, stone monuments, wide paved streets, canals, and long aqueducts. It was connected to the surrounding mainland by causeways.

The Aztecs attributed the Earth's tremors that often struck México-Tenochtitlan to the god Tepeyollotl, who ruled over earthquakes, echoes, dark caves and jaguars. As a nation professing the cult of the Sun, they believed that the end of the world would come in the form of an earthquake and destruction of the Sun. The Aztecs left behind records of earthquakes in the fifteenth century, and supposedly were able to predict them using the Sun stone. However, it is not known how they did this.

Hernán Cortés In March 1519, the Spanish conquistador Hernán Cortés (1485–1547) landed with Spanish troops in Veracruz, about 400 km east of the present-day capital of Mexico. The motivation for the expedition to this part of North America was gold, which was said to be abundant among the Aztecs.

On the way to Veracruz, Cortés stopped in the Yucatán, where a young indigenous woman, the educated and beautiful daughter of a Yucatán merchant, joined his crew as an interpreter. She later became the mother of Cortés' son Martín. Malintzin, "La Malinche" or Doña Marina after her Christian

Fig. 7.1 México-Tenochtitlan—the legendary wealthy capital of the Aztec empire as imagined by the artist. The city was founded in 1325 on an island in Lake Texcoco in the Valley of Mexico. The ruins of the city can be found in the historic centre of the current capital. © Ladislav Csurma, 2023. All rights reserved

baptism, became an iconic figure in Mexican history, personifying the betrayal and sell-out of her homeland to the European conquerors. The term 'malinchismo' still denotes the act of showing affection for the foreign and contempt for one's own.

When the Aztecs learned about the unknown beings seen near Veracruz, they believed it was the mythical return of their god Quetzalcoatl. On 8 November 1519, the Spaniards arrived in the city of México-Tenochtitlan (Fig. 7.1). They saw a city of unprecedented wealth, magnificent architecture, and above all, a vast amount of gold. The Aztec ruler Moctezuma II (reigned 1502/1503–1520) welcomed the Spaniards and exchanged gifts with Cortés. The city's inhabitants experienced the arrival of the Spaniards as an unbelievable dream, but at the same time, they were filled with fear, a premonition that their world would soon be turned upside down.

Everything began to change significantly just a few days after their arrival. Horrified by the bloody religious practices and fearing for their own lives, the Spaniards imprisoned Moctezuma. At the same time, they could not suppress their desire for Aztec gold, and the Aztecs soon grew impatient with their constant demands, interference in religious practices, and attempts to replace their idols with Christian symbols.

Moctezuma, perceived as an insufficiently assertive ruler, was ultimately rejected by his own people, and the city closed itself off. The Spaniards were expelled from the island, but several hundred were imprisoned in the city and subsequently sacrificed in mass religious rituals.

With the help of the indigenous population, who had been subjugated by the Aztecs, Cortés conquered the city of México-Tenochtitlan in August of 1521 and began transforming it into Spanish Mexico.

Cortés wanted to build a new Spanish centre on the island. Other key members of his crew leaned towards a location closer to the mountains and pastures, where they could raise cattle as they were accustomed to, and where the terrain provided natural water regulation. The decision to build the capital on a lake island was purely Cortés'—reportedly, it allowed for rapid communication with the population on the lake shores. Another reason why Cortés chose to rebuild México-Tenochtitlan could have been the symbolic significance of this step—he sought to erase any memory of the past and thereby prevent any attempts at rebellion.

During the following centuries of Spanish colonization and the city's expansion, the Spaniards faced the consequences of the city's island location and the surrounding lake. The rise in water levels during summer rains caused flooding. Deforestation along the shores and neglect of the original Aztec drainage and irrigation system exacerbated the flooding situation. Mud released due to deforestation settled at the bottom, and the lake became increasingly shallow.

At the beginning of the seventeenth century, they embarked on the construction of the Desagüe, a hydraulic drainage system that became one of the most complex engineering feats of the pre-industrial era. However, it was not completed until the nineteenth century. Until the system reached its peak condition, during certain periods, it was neglected and failed to such an extent that people began to move en masse from the island to the more stable shores of Lake Texcoco. In 1630, there was even a proposal for the official relocation of the city from the island to the mainland around the lake.

In the eighteenth century, the lake level receded, creating conditions for the expansion of the city into areas that were initially submerged by lake water. In the nineteenth century, the drainage project of the main city of Mexico was driven by increased awareness of the need for hygiene and the health risks posed by the stagnant waters. However, floods persisted until the twentieth century when the territory of the capital city was drastically cleared of remnants of water through landfill.

Today, only small remnants of Lake Texcoco survive around the city in the form of ponds and salt marshes, which serve as a refuge for the last remnants of endangered fauna and flora (such as the endemic Mexican axolotl). These are to be maintained as an ecological park under the protection of the Mexican government.

What Is the Capital of Mexico Built on? Geological and Geotechnical Views

The Valley of Mexico is the southern and volcanically active part of the volcanic-tectonic region known as the Trans-Mexican Volcanic Belt, which extends from the eastern to the western coast of Mexico. This region is composed of older Tertiary volcanic rocks overlaying Cretaceous limestones. On the surface of these formations in the Valley of Mexico, there are younger Tertiary volcanic rocks covered by a layer of tuffs or sands, gravels, and recent lava flows, approximately 100 m thick. The average thickness of rocks above the Cretaceous limestones is approximately 2 km.

This bedrock characterises the hilly area of the Valley of Mexico and parts of the capital, and can be considered solid from a geotechnical and seismological point of view. Within the area bounded by this bedrock, a sedimentary basin can be distinguished in which a significant part of the upper layer is made up of unconsolidated lacustrine sediments and alluvial deposits. These are mainly composed of soft sands and clays. Quite a large part of the present-day capital is built on these sediments. Their thickness varies from about 10 m to 100 m, depending on the location. The approximate average thickness is 50 m. The water table is in the depth of about 2 m over most of the sedimentary basin. Therefore, a large part of the city lies on a relatively thick layer of soft, water-saturated lake sediments.

It is not surprising that a major problem in the capital is that large and heavy buildings built in the past are sinking into the soft lake sediments. Many important buildings are in danger of collapse. That was also the case with the Metropolitan Cathedral of the Assumption of the Most Blessed Virgin Mary into Heaven in the Plaza de la Constitución. It is the largest Roman Catholic Baroque cathedral in the world and the seat of the Archdiocese of the Primada de México. Extensive renovations in the 1990s stabilized, and thus saved, the cathedral.

The settling of lake sediments and the sinking of buildings is a major problem. However, not the biggest Mexico's capital is facing.

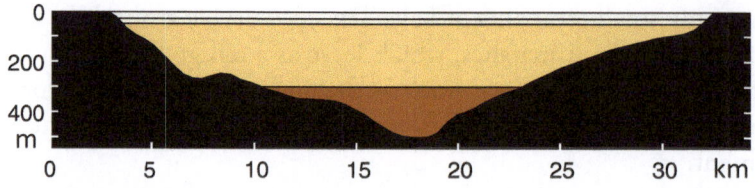

Fig. 7.2 Vertical cross-section of the simplified structural model beneath the territory of the capital city of Mexico, where catastrophic damages occurred due to an earthquake in 1985. Using this model, it was only recently possible to explain anomalous seismic motion on the territory of the Mexican capital. The different colours of sediment layers represent various values of the speed of seismic S-waves. Based on Cruz-Atienza et al. (2016), © The Authors, 2016. CC-BY-4.0 license

On average, in the upper 50 m thick layer of lake sediments, the speed of S-waves (seismic shear waves) ranges from 30 m/s to 100 m/s. The speed of P-waves (seismic compressional waves) is approximately in the interval from 800 m/s to 1200 m/s. The value of the S-wave speed and the value of the ratio of the P-wave and S-wave speeds are extreme. S-wave speed low, speed ratio high.

In a simplified model, beneath the upper layer of extreme unconsolidated lake sediments, a layer with an S-wave speed of approximately 400 m/s (that is four times larger) can be considered to a depth of approximately 300 m. Beneath that, there is a layer to a depth of approximately 500 m with the S-wave speed of about 800 m/s. These sediments form the interior of a sedimentary basin in a substantially stiffer basement, where the S-wave speed is approximately 1600 m/s. A vertical cross-section of such a sedimentary basin is shown in the figure (Fig. 7.2).

These are not useless "technical" details. On the contrary! They are important data that, only recently, after decades of effort, made it possible to explain the anomalous seismic motion during the 1985 earthquake on the territory of the Mexican capital. We will explain further below.

Earthquakes in Mexico

Based on found historical documents or seismometrically recorded earthquakes, five earthquakes with a magnitude of 8 or greater have occurred in Mexico or immediately off its coast since the second half of the eighteenth century. In the same period, there have also been 59 earthquakes with magnitudes between 7.0 and 7.9. This is significant seismic activity, and Mexico is in fact one of the most seismically active regions in the world. This is the result of the complex relative motion of three lithospheric plates—the North American Plate, the Pacific Plate and the Cocos Plate.

Let's mention particularly the earthquake of 1475 in México-Tenochtitlan. The only mention of this earthquake has so far been found in the so-called Aubin Codex (named after the French collector and Americanist J. M. A. Aubin). The codex was written in the Nahuatl language and dated between 1576 and 1607. It describes the history of the Aztecs starting with their migration from Aztlan up until the arrival of the Spanish conquistadors and the conquest of the capital city of México-Tenochtitlan. The codex was referenced by the Franciscan monk Juan de Torquemada (1562–1624) in his work *Monarquía indiana*, which is one of the fundamental texts describing the history of Mexico. He conveyed to subsequent generations that the earthquake 'was so strong that not only many houses fell, but also mountains and mountain ranges shattered and broke in many places.' The earthquake likely also caused a tsunami in Lake Texcoco. According to current estimates, the earthquake had a magnitude of about 7.5 and occurred in present-day Cuajimalpe due to local seismic faults.

Very simply, the epicentres of localised earthquakes can be found virtually all over Mexico. However, earthquakes occur mainly in two seismotectonic zones.

The majority of earthquakes and the largest earthquakes occur in the subduction zone beneath the southern part of Mexico. The subduction zone extends along the west coast of Mexico from approximately Isla Maria Madre to the border of Costa Rica and Panama. In the subduction zone, the Cocos lithospheric plate underlies the North American plate at an angle of about 14 degrees. This subduction is not simple: in the northwestern part of the zone it is at a rate of about 68 mm/year, and at the southern end of the zone it is at a rate of up to 78 mm/year. As noted in the previous chapter, these are large rates. The subducting of one plate beneath the other causes not only large earthquakes but also occasional volcanic eruptions. It also forms the deep Middle America Trench along the south-west coast of Mexico. All 5 known earthquakes in Mexico with a magnitude of 8 or greater have occurred in this subduction zone.

According to current knowledge, the largest earthquake occurred on 28 March 1787, at 11:30 AM local time, known as the San Sixto (Guerrero-Oaxaca) earthquake. It is estimated that the moment magnitude could have been around 8.5, and the earthquake was caused by a relative displacement between the Cocos Plate and the North American Plate along approximately 450 km off the southern coast of Mexico. The earthquake destroyed many buildings in the city of Oaxaca and also damaged buildings in the capital city.

Additionally, the earthquake triggered a tsunami, which, according to estimates, reached a height of 18.5 m.

The second most important seismotectonic zone is the contact of the North American plate with the Pacific plate. The Pacific plate moves relative to the North American plate in a northwest direction and, conversely, the North American plate moves in a southeast direction relative to the Pacific plate. And with it the territory west of the Gulf of California, including the Baja California peninsula. The rate of the relative motion is approximately 5 cm/year.

On 28 July 1957, a magnitude 7.6 earthquake occurred in the subduction zone with an epicentre in the Mexican state of Guerrero, about 260 km south of the Mexican capital. Despite this distance, the earthquake caused the greatest and unprecedented damage in the capital area. Many buildings collapsed. Later investigations revealed that all the buildings had serious defects. Despite these flaws, it was clear to experts that the seismic motion in the capital was peculiar—strong shaking lasted up to 90 s. The location of the greatest damage indicated the strong influence of local geological conditions.

The earthquake significantly damaged 90% of the buildings in Chilpancingo—the capital city of Guerrero state. Significant damages were also reported in other cities within Guerrero state. However, these damages were not surprising considering the locations of the cities. The real surprise came from the significant damages in a considerably distant capital city. In 1957, no one anticipated that a truly unimaginable and shocking surprise was yet to unfold, getting ready even further away from the Mexican capital.

Before the earthquake in 1957, seismic vulnerability was not adequately addressed in building regulations. After the 1957 earthquake, building codes were tightened. According to the version from 1976, the territory of the Mexican capital was divided into three zones: the original lake zone, the transitional zone, and the solid bedrock zone. The building code differentiated between ordinary structures and very important structures.

However, in actual construction after 1957, certain aspects, such as the distances between neighbouring buildings, were not adhered to. The low quality of sand and gravel often resulted in concrete being significantly lighter than what the building code assumed. As a result, the elastic modulus was lower than required. Similarly, the ratio of tensile stress to compressive stress was affected, leading to a reduction in material lifespan due to repeated cyclic loading. Material shrinkage was exceptionally high. The consequences of using slow-setting cement also posed a problem.

19 September 1985 Earthquake and an Unprecedented Catastrophe in the Mexican Capital

Even today, the events of 19 September 1985 are hard to believe for laymen and seismologists alike. Despite the 1957 earthquake, and despite the fact that the inhabitants of the capital were used to earthquakes of magnitude 6 every few years.

At 7:17 AM local time, an Mw 8.0 earthquake began at a depth of about 28 km, about 110 km northwest of the city of Zihuatanejo (near the famous tourist destination of Ixtapa), in a subduction zone at the contact between the North American and Cocos plates. The epicentre was inland, several tens of kilometres from the coast.

Macroseismic effects were observed in an area of 825,000 km². Approximately 20 million people overall experienced the earthquake. The maximum macroseismic intensity, IX on the 12-degree Modified Mercalli Intensity Scale (MM), was observed in the coastal cities of Lázaro Cárdenas, La Unión, and Ixtapa, located approximately 45, 74, and 114 km from the epicentre, respectively. However, effects were also felt in Ciudad Guzmán, approximately 190 km from the epicentre, and in the capital city at distances of over 350 km from the epicentre. The earthquake also caused damage in the states of Colima, Guerrero, Mexico, Michoacan, Morelos, and in parts of Veracruz and Jalisco.

According to official sources, more than 9500 people lost their lives as a result of the earthquake. The public did not believe this figure, and according to many estimates, more than 40,000 were dead. Fortunately, many buildings, especially schools and administrative buildings, in the capital were not full of people at the time of the earthquake. If the earthquake had occurred 2–3 h later, there would have been even more victims.

According to official sources, around 250,000 people lost their homes. The public believed in a more realistic estimate of approximately 750,000 people without a roof over their heads.

In the capital of Mexico, at a distance of 358–366 km (!) from the epicentre of the earthquake, 412 buildings collapsed and another 3124 were seriously damaged (Fig. 7.3).

The collapse of the Juárez Hospital in the Mexican capital was tragic and particularly sad. At the time of the earthquake, 430 beds were occupied and the morning shift change was in progress. The rescue operation was complicated because ambulances were parked inside the collapsed building. There

Fig. 7.3 Ruins of a 15-storey residential building in the territory of Mexico City. USGS—public domain

were neither the facilities nor sufficient staff to provide urgent care to the injured. As a result of the hospital collapse, 561 people died. 188 victims were never identified. Others were identified by family members or through dental records. However, the unidentified victims do not include the human body parts recovered.

Also linked to the tragic event is the story of the 'miracle babies' who were rescued from the ruins of the buildings after 7 days without food, warmth or human contact. Sadly, their mothers did not survive the earthquake.

The Secretariat of Communications and Transport, with its satellite tower, also collapsed. The capital was in danger of being cut off from communication with the rest of the world. In a disaster situation, this would have significantly reduced the chances of the capital getting the help it needed.

The subway trains stopped suddenly and thousands of people scattered through the tunnels looking for a way out. Almost six million people suddenly lost access to drinking water in their homes. 40% of the population remained without electricity and 70% without a telephone connection.

Efforts to restore transport, electrical networks and telecommunication connections were repeatedly thwarted by earthquake aftershocks.

The capital lost three quarters of the total original number of hospital beds. 200,000 people lost their jobs.

The earthquake caused damage estimated at 5 billion US dollars (in current terms, up to 11 billion US dollars). The earthquake was a humanitarian and economic disaster for the entire Mexico. An unusual example of its consequences is that even 20 years after the earthquake, 80 families still waited for permanent housing as compensation for collapsed apartment buildings.

The earthquake also triggered numerous landslides and rockfalls. It also caused a tsunami, reaching a maximum height of 3 m in Zihuatanejo. In the port of Hilo on the island of Hawaii, it was only 21 cm.

Seismic Gap

The earthquake occurred in a zone known as a "seismic gap"—there had been no earthquakes there for decades. Perhaps earlier, in 1911, but even that is not certain. However, given the general subduction of the Cocos Plate beneath the North American Plate, many assumed a great potential for a major earthquake in this zone.

The relative motion between the plates occurred over an area of about 170 × 50 km², with some estimates as large as 180 × 80 km². The rupture propagated for about 1 min on the fault at an average velocity of about 2.8 km/s. On average, the plates were relatively displaced by 230 cm. In the epicentral region, the Earth's surface was lifted by about 1 m and displaced 2 m in a south direction. On the ocean floor near the coast, there was uplift on the North American plate side and subsidence on the Cocos plate side. This seafloor movement caused the tsunami.

The maximum acceleration recorded at the nearest seismic station to the epicentre, Caleta de Campos, was 138 cm/s², 141 cm/s², and 89 cm/s² in the north, west, and vertical directions, respectively.

An interesting point is that the maximum accelerations in the epicentral area were not as large as expected for earthquakes with a moment magnitude of 8.

Based on the analysis of acceleration records in the area above the fault, maximum seismic motion was found at periods between 2 and 3 s.

What Happened in the Capital?

Above all, it is necessary to emphasize that large parts of the city had no damage. At that time, the capital had over 18 million inhabitants and approximately 800,000 buildings. When we compare these numbers with the number of casualties and the number of collapsed and seriously damaged buildings, we see that strong seismic motions fortunately only damaged a relatively small part of the city. However, this does not change the fact that the absolute numbers of casualties, homeless people, and collapsed and seriously damaged buildings are terrifying.

It is very interesting to compare the maximum seismic accelerations recorded by seismic stations in the capital. Stations on a firmer bedrock, in hilly areas, and in transitional zones recorded a maximum of 49 cm/s² (this value was recorded on the north-south component by the Teacalco station on volcanic tuff). However, the situation was substantially different on the surface of soft lake sediments. The SCT (Secretaría de Comunicaciones y Transporters) station recorded a maximum acceleration of 168 cm/s², i.e., 0.17 g, on the east-west component. The SCT station is located on the surface of soft lake sediment layers. It is worth noting that near the epicentre, a maximum acceleration did not reach more than 141 cm/s²! It's as if the seismic waves were not damped at all when propagating over such a great distance but rather intensified! (Fig. 7.4).

The value of the maximum acceleration on the surface of soft lake sediments wouldn't cause such damage if the seismic motion didn't have a

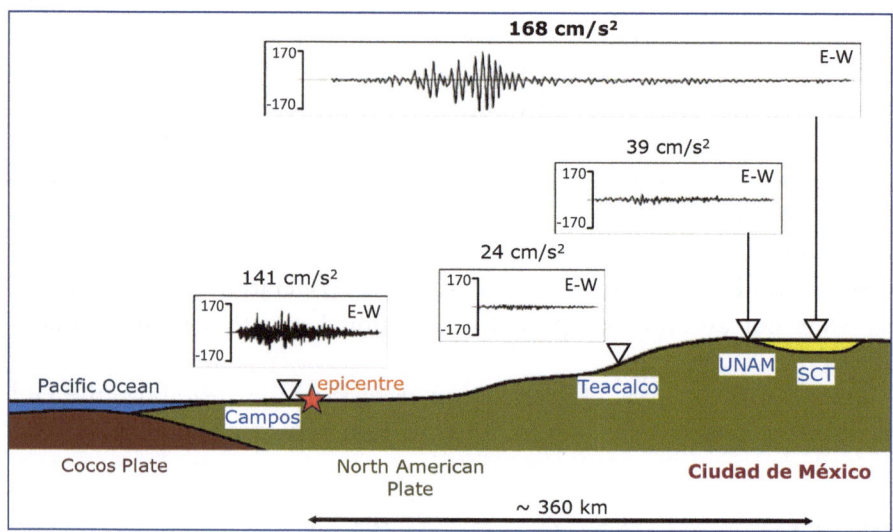

Fig. 7.4 The 1985 earthquake originated in the zone of the Cocos Lithospheric Plate subducting beneath the North American Plate. The acceleration of the seismic motion in the east-west (E-W) component in the territory of the capital was greater than near the epicentre. It is one of the most prominent cases of anomalous seismic motion due to local conditions in part of the territory of the Mexican capital

sufficiently long duration. Some seismic records show up to 15 cycles of motion with significant amplitudes at periods close to 2 s. These periods are where seismic motion is most "enhanced" (amplified and prolonged) by the local soil conditions—layers of soft lake sediments. Indeed, at these periods, the acceleration at the SCT station was approximately 10 times greater than that at the UNAM (Universidad Nacional Autónoma de México) station on firmer ground.

However, not all buildings in the lake sediment zone were damaged. Buildings with less than 5 storeys and modern buildings with more than 30 storeys were not affected at all or suffered very little damage. However, most damaged buildings had between 5 and 20 storeys (Fig. 7.5).

As an approximate rule of thumb for North-American buildings, the period of the fundamental mode of the building's natural vibrations (the

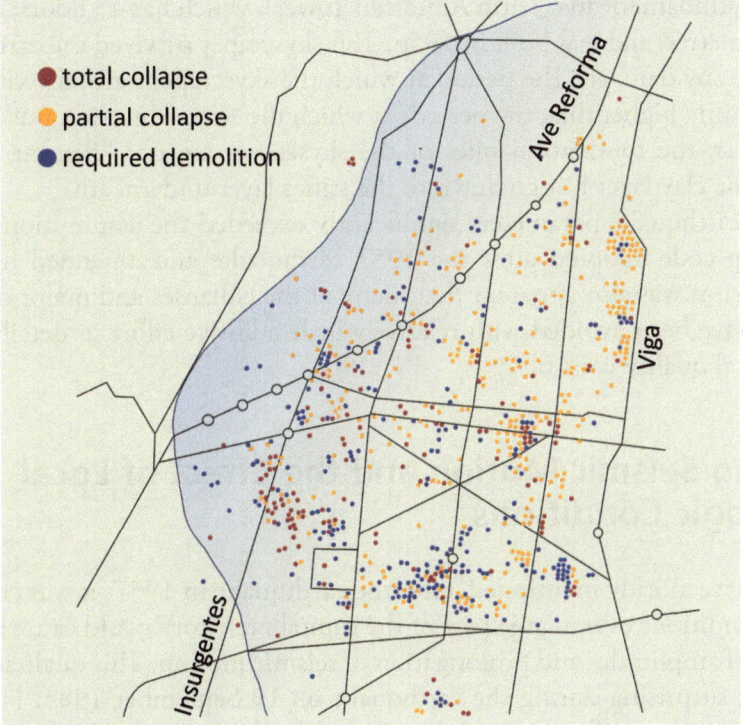

Fig. 7.5 Map of the damage caused by the earthquake of 19 September 1985 in the Mexican capital at a distance of more than 350 km from the earthquake's epicentre, which was on the west coast of Mexico. All the serious damage was confined to the sites of the original Lake Texcoco. The blue line marks the boundary of the lake before it was artificially drained. Based on Stone et al. (1987), © U.S. Department of Commerce, 1987. All rights reserved

fundamental period at which the building 'likes' to vibrate) with N floors is N/10 s. (In Europe it is closer to N/20.) In the case of the buildings collapsed or significantly damaged in the Mexican capital, the period estimated in this way was close to the fundamental resonant period of the local soil structure (about 2 s, i.e. the period at which the local soil structure 'likes' to vibrate).

It is very likely, therefore, that the amplification of seismic motion by layers of lake sediments (increasing amplitudes and prolonging duration due to the specific soil conditions described above), natural vibrations of buildings, and subsequent amplification of seismic motion by vibrating buildings, combined with the structural properties and construction deficiencies of buildings, caused the collapse or significant damage to buildings. The destruction was also contributed to in some places by "soil collapse" due to strong cyclic loading.

It is worth mentioning the case of the oldest skyscraper in the capital, the Torre Latinoamericana (Latin American Tower), which has 45 floors, a height of 183 metres, and was built in 1956. This skyscraper survived the earthquake without any damage. The period at which the skyscraper naturally vibrates is significantly higher than the periods at which the seismic motion was intense. Moreover, the foundation piles of the skyscraper are not "floating" in the lacustrine clay layer but go down to the stiffer layer underneath.

The earthquake parameters significantly exceeded the assumptions of the building code adopted after the 1957 earthquake and amended in 1976. However, it was also apparent that many of the collapses and major damages could have been avoided with relatively little additive effort in detailing and structural quality control.

Strong Seismic Motion and the Effect of Local Geologic Conditions

As we have already mentioned, after the earthquake in 1957, it was clear that local conditions over a large part of the capital's territory could cause amplification of amplitudes and prolongation of seismic motion. This qualitative fact was not surprising during the earthquake on 19 September 1985. However, the extent of amplification and the scale of damage were shocking.

The problem and unprecedented challenge for seismologists and seismic engineers were, therefore, to quantitatively explain the intense and prolonged seismic motion and the behaviour of buildings.

Considering the earthquake in 1957, seismologists immediately suspected that the anomalous seismic motion on the surface of lake sediments was caused by a combination of three factors: the source itself generating strong seismic waves with periods of 2–3 s, the propagation of waves from the source to the capital, which did not attenuate these seismic waves as expected at such a distance, and finally the amplification of waves in the layers of lake sediments.

Quantitatively, however, it has been unclear for several decades how the long duration of seismic motion, specifically surface seismic waves, in soft lake sediment layers can be explained. Attempts to numerically model this motion have run into the problem of strong attenuation in the sediments. However, a very plausible explanation seems to have been reached by a team of UNAM seismologists led by Professor Víctor Manuel Cruz-Atienza. Unlike previous attempts, the team considered in their numerical modelling not only the surface layers of the lake sediments, but the structure of the bedrock at all depths. We have described the model in simplified terms above. The numerical modelling showed that even in the realistic model, the so-called fundamental mode of surface seismic waves is indeed strongly attenuated over short distances. However, and this is a new finding, the first higher mode of surface seismic waves, strongly supported by the bedrock beneath the shallower surface layers, can propagate at periods close to 2 s throughout the sedimentary basin with significant amplitudes.

Wrong Political Decisions

In the 1980s, Mexico was grappling with high foreign debts. Decades of remarkable economic and political stability prior to the 1980s, known as the "Mexican miracle," were already a distant past.

However, despite this, the Institutional Revolutionary Party (Partido Revolucionario Institucional, PRI) and President Miguel de la Madrid (in office from 1982 to 1988) refused foreign aid after the earthquake. A ban on the dissemination of information was imposed. No one trusted official figures on the number of casualties and people left homeless. The real numbers were significantly lowered on purpose. The government's immediate decisions and actions were uncoordinated. The government ordered the evacuation of expensive equipment from damaged factories before evacuating injured city residents. President de la Madrid declared a three-day mourning period, during which he made no official public appearances. His subsequent insensitive and authoritarian statements in response to the tragedy led to accusations of incompetence.

The government became the target of harsh social criticism for its irresponsible response to the humanitarian crisis caused by the earthquake. The PRI was also accused of favouring construction companies based on personal relationships before the earthquake, resulting in poor quality construction. Many of these buildings collapsed in the earthquake, needlessly killing thousands of people.

The PRI, which has been in power since the 1930s, completely failed to deal with the crisis and was reluctant to implement the measures set out in the national contingency plan. This, too, contributed to the loss of confidence in the party.

Internally, however, a wave of great solidarity emerged within society. Among young people, an unofficial group of volunteers was formed, who, without adequate training and safety equipment, risked their lives in unstable ruins in search of survivors. They also distinguished themselves by rescuing newborn babies after the collapse of the Juárez hospital. In 1986, they became an official multinational group, which is now specially trained and leads rescue operations in earthquake-affected areas worldwide. The spontaneous mobilization of volunteers has proven to be an effective tool in addressing acute problems. Instead of supporting this initiative, however, the government hindered it in an attempt to gain control.

From regional political groups and the so-called "damnificados," people who lost most of their property due to the earthquake, a strong political opposition against the PRI began to form. They collectively acted as a unified coordination committee of disaster victims. The committee forced President de la Madrid to ensure the construction of 100,000 homes by 1987.

Under public pressure, the government was eventually forced to accept foreign aid—60 states and numerous non-governmental organizations provided humanitarian support to Mexico.

Reputation of the PRI, which had been on decline for years, was permanently damaged due to its inadequate response to the earthquake. After nearly 60 years of unchallenged rule, the party faced defeat in the 1988 elections. It tried to prevent defeat by attempting to manipulate the election results. Society responded with several turbulent weeks of unrest, ultimately leading to the complete paralysis of the government for a whole week.

Future Earthquakes

Mexico and its capital will continue to face strong and threatening earthquakes in the future. The earthquake on 19 September 1985, released tension in the quiet seismic zone in the states of Michoacán and Guerrero. However, to the east of the state of Guerrero, approximately between the epicentral area of the 1985 earthquake and Acapulco, there is another quiet zone. It is not excluded that a large earthquake may occur there at some point in the future.

Every year on 19 September, a drill focused on operations and evacuation is held in order to get ready for the next major earthquake. Coincidentally, on 19 September 2017, about 2 h after the drill ended, the cities of Mexico City and Puebla were hit by an earthquake with a moment magnitude of 7.1. Also in 2022, on the day of the exercise, a 7.7 magnitude earthquake struck the states of Michoacán and Colima on the west coast of Mexico.

After the earthquake on 19 September 1985, building regulations were tightened. However, these regulations do not anticipate earthquakes with a magnitude greater than 8. This may be a similar mistake to what was made in Japan before the catastrophic earthquake and tsunami in March 2011.

In the 1990s, seismic stations were installed along the subduction zone as part of the Mexican Seismic Alert System (Sistema de Alerta Sísmica Mexicano, SASMEX). The aim is to utilize the time needed for intense seismic waves to reach the capital. Depending on the location of epicentre on the southern coast, seismic waves reach the capital after approximately 60–90 s. However, the time needed to locate the earthquake and estimate its magnitude is less than 10 s. Thus, after locating an earthquake with a magnitude higher than 6.0 in the subduction zone on the southern coast, there is enough time for electronic warnings to be sent to the capital and for taking important measures in the capital. The warning allows timely shutdowns of trains, metros, hospital operations, and taking other actions. People can prepare for aftershocks.

This early warning system was the first such operational system and has already proven to be effective.

8

Ancient Witnesses of Medieval Earthquakes

As we saw in the previous chapter, it was specific local geological conditions on the territory of the capital of Mexico that caused shocking damage at a distance of more than 350 km from the epicentre of the earthquake.

In this chapter, we will zoom in on the surprising local effects due to medieval earthquakes on the territory of Rome.

Accademia Nazionale dei Lincei

In October 1993, the unique scientific conference *Terremoti e civiltá abitative: Nuove discipline e applicazioni. Dieci anni di ricerche* (Earthquakes and civilization settlements: New disciplines and applications. Ten years of research) took place in the headquarters of the learned society of Italy, the Accademia Nazionale dei Lincei, founded in 1603. At the conference, 50 papers were presented, which resulted from research into the relationship between historical earthquakes and settlements. Very interesting and surprising findings were related to ancient buildings that have survived to this day and become unique witnesses of medieval earthquakes. Perhaps the most famous such object is the Roman Colosseum.

One year before the conference, Professor Renato Funiciello, Italian geologist and Vice President of the National Institute of Geophysics in Rome (now Istituto Nazionale di Geofisica e Vulcanologia) and seismologist Dr. Antonio Rovelli asked the first author of this book to perform numerical simulations of the seismic response under the Roman Colosseum and invited him to

© The Author(s), under exclusive license to Springer Nature Switzerland AG 2024
P. Moczo et al., *Earthquakes*, Springer Praxis Books,
https://doi.org/10.1007/978-3-031-64707-9_8

present the results at the conference. The results were then published in the *Annali di Geofisica* in 1995, in the issue devoted to the conference papers.

After the conference, research on the effects of earthquakes on preserved ancient buildings continued. Italian seismologists Enzo Boschi, Antonio Rovelli and their colleagues published an article in 1995 in the journal *Bulletin of the Seismological Society of America* about ancient Roman columns.

The Mystery of the Roman Colosseum

Roman Colosseum

The Colosseum in Rome is the largest and best preserved of the still standing Roman amphitheatres. Its foundation stone was laid in 72 AD by order of Emperor Vespasian (reigned 69–79 AD) of the Flavian dynasty as a celebration of the suppression of the Jewish uprising and as a sign of generosity to the Roman people. Construction was completed by his sons, Emperors Titus (reigned 79–81 AD) and Domitian (reigned 81–96 AD). 100,000 m^3 of stone was imported for the construction and 60,000 Jewish captives—slaves—participated in it.

The Colosseum was built on the central part of Nero's architectural masterpiece Domus Aurea, where the artificial lake Stagnus Neronis was supposed to extend. It was named after the "colossal" bronze statue of Emperor Nero (reigned 54–68 AD) standing nearby.

The construction of the Colosseum was part of Vespasian's project to revitalize Rome. The monumental building was supposed to become the successor of the Roman Forum, whose cultural tradition had been built since the time of Romulus, the legendary founder of Rome and its first king.

After almost a decade of construction work, the Colosseum was inaugurated by Emperor Titus in 80 AD. The celebrations lasted 100 days and the silver sesterces minted that year bore the image of Rome's newest landmark. As many as 50,000 people could watch the gladiatorial death matches, imitations of famous battles, or the fierce *venationes* (a confrontation between man and beast) at the same time. Between the construction of the Colosseum and the ban on fights in the fifth century, approximately half a million people lost their lives in the Colosseum.

As a result of the urbanization processes of the High and Late Middle Ages, the Colosseum underwent significant functional changes. Housing and craft units were set up in the areas under the auditorium, a place was also found for a Christian chapel and the arena became a cemetery. The medieval papacy tried to turn the Colosseum into a place of martyrdom and make a relic from the sand on which Christian martyrs bled. Since 1998, the Colosseum has been a world symbol of the fight against the death penalty.

Even those who have not seen the Colosseum with their own eyes know from photos and media that about 50% of the outer wall on the south side is missing, while the north wall is practically intact.

The current state of the Colosseum was caused by various factors. Fires, lightning, subsidence and earthquakes also contributed. Many human interventions, damage or modifications probably made it impossible to distinguish some damages caused by earthquakes. The marble stands and seats were taken by the Renaissance popes and the Roman aristocracy. They needed them in new palaces and temples. The interior of the Colosseum was rather wildly modified.

Earthquakes in Italy

The catalogue of Italian earthquakes is exceptional in the world: it covers well a period of more than 2000 years. It also includes several large earthquakes (with an estimated magnitude of 7) in the central Apennines, which were also felt in Rome, 100 km away. In Rome, they caused damage up to the seventh degree of the twelve-degree Modified Mercalli Macroseismic Intensity Scale.

The 29 April 801 Apennine earthquake, 25 January 1348 Friuli earthquake and 14 January 1703 Apennine earthquake caused enormous damage in the city of Rome. Although reliable and detailed information on the damage caused by individual earthquakes is lacking, several inscriptions and chronicles mention the reconstructions and restoration of the Colosseum commissioned to repair earthquake damage.

None of this, however, leads to an explanation of why the outer north wall is undamaged and the entire northern section almost intact, but the outer south wall is missing, and the southern section is visibly more damaged … (Fig. 8.1)

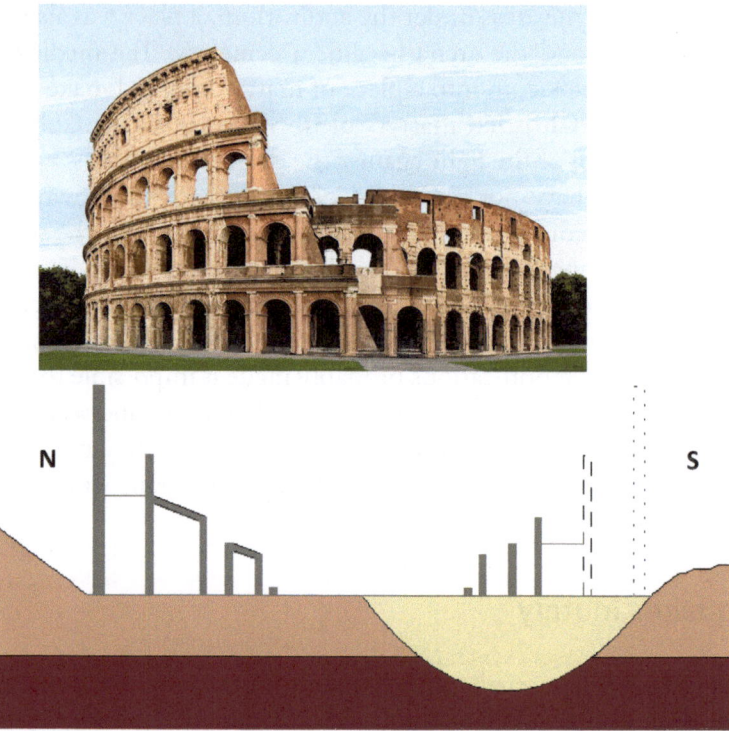

Fig. 8.1 A view of the famous Colosseum in Rome. The northern part of the outer wall is very well preserved, the southern part is completely missing. A paradox or even a mystery that has not been explained for a long time. The lower part of the image: a vertical section of the underlying geological structure in the north-south direction. The southern part of the Colosseum was built on the surface of the sedimentary fill of the former tributary of the Tiber River. From the point of view of statics, a manageable situation, from the point of view of the dynamic behaviour of the subsoil and the Colosseum during an earthquake, a big problem. Reprinted with permission from Moczo et al. (2023), © GRADA Slovakia s.r.o., 2023. All rights reserved

Surprising Finding

Professor Renato Funiciello found that while the northern part of the Colosseum lies on a near-horizontal layer of consolidated continental sediments from the Pleistocene (a geological period that began about 2.6 million years ago and ended about 11,700 years ago), the southern part rests on a sedimentary fill of a former tributary of the Tiber River. This means there is a confined sedimentary valley beneath the southern part of the Colosseum. And this is the fundamental difference between the northern and southern sides. The stability and survival of structures and buildings depends not only on good statics, but also on how the building behaves, for example vibrates, due to external factors. This mechanical vibration is, of course, highly dependent on how the immediate subsoil vibrates.

The Southern Part of the Colosseum

The sensitivity of the southern part of the Colosseum to earthquakes became apparent already in 443 and 484. The earthquakes damaged the stage and arena. The Colosseum was also damaged by earthquakes in the winter of 1703, which had epicentres in the region of Umbria-Marche and L'Aquila. It was not until the middle of the eighteenth century that Pope Pius VII (1742–1823, pontificate from 1800–1823) began the first reconstructions.

The sedimentary fill of the former tributary of the Tiber River forms a relatively deep valley. Deep in this case is a relative term: it is the ratio of the maximum depth to the maximum width.

The sediments in the valley are seismologically soft compared to the material around and beneath the valley: shear waves propagate 2–4 times slower in the valley than in the horizontal layer and 4–8 times slower than in the bedrock beneath the deepest part of the valley. What does this mean? The valley becomes a trap for seismic waves that enter the valley. In this trap, the waves slow down and bounce many times between the surface and the sediment-bedrock interface. Only a fraction of the energy that has entered the trap will return to the vicinity of the valley due to the high velocity contrast at the sediment-bedrock interface.

The interference of multiply reflected waves can cause a significant amplification of the oscillatory motion at some frequencies in the valley and on its surface. If the valley was shallow, surface waves would mainly arise in the valley, which would propagate in a horizontal direction. However, if the valley is deep (which also depends on the contrast of the wave speeds between the valley and its surroundings), something even more interesting and dramatic occurs: the resonance of the entire valley. That is very similar to what happens throughout the Earth when the Earth's own oscillations arise. (Recall the paragraph on anomalous seismic motions in the brief introduction to earthquakes.)

Amplification of amplitudes at specific frequencies can be very significant. Therefore, in the valley below the southern part of the Colosseum, at these frequencies (in the range 1–4 Hz), the ground motion is significantly stronger (even by one order) than in the vicinity of the valley (Fig. 8.2).

Such resonant amplification of seismic ground motion in the sedimentary valley beneath the southern wall of the Colosseum is the most likely

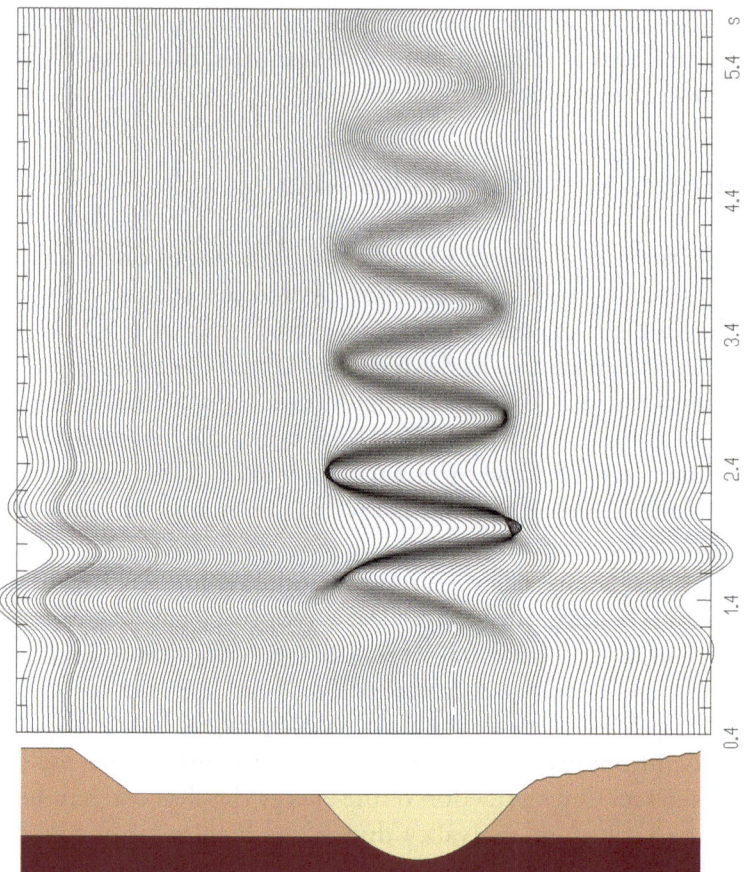

Fig. 8.2 Numerically simulated vibration of the bedrock beneath the Colosseum in Rome. Each curve represents a simulated time record (vertical axis corresponds to seconds) of the vibration at a given point on the Earth's surface at the frequency of the so-called fundamental mode of resonance of the sedimentary valley beneath the southern part of the Colosseum. It can be seen that the oscillation of the valley is resonantly amplified compared to its surroundings. Reprinted with permission from Moczo et al. (2023), © GRADA Slovakia s.r.o., 2023. All rights reserved

explanation so far for the unusual sensitivity of the southern part of the Colosseum to earthquakes. And that the southern part of the outer wall is missing: after being damaged by one or more earthquakes, the material of the wall was taken away by those who used it for their own constructions.

A supporting argument for the very different ground motion beneath the northern and southern parts of the Colosseum are also the results of measurements of seismic noise at the base of the Colosseum and in its upper part. As already mentioned in the chapter Earthquakes—a Short Introduction for Almost Everyone, seismic noise is a continuous weak vibration of the Earth due to a multitude of different waves induced by thousands of different natural and technogenic sources. Significant amplifications in seismic noise were found at the same frequency interval (mainly between 1 and 2 Hz) that was indicated by numerical simulations.

In section on Amplified Seismic Motion we briefly characterized the global resonance of the sedimentary valley. What we describe here is the case of such phenomenon. The global resonance was identified by American seismologist Dr. Brian Tucker and his colleagues who performed special measurements of earthquake motion in deep sedimentary valleys in Tajikistan in 1984. Based on this observation, French seismologists Dr. Pierre-Yves Bard and Dr. Michel Bouchon developed theoretical explanation and numerical modelling of the global resonance.

Colosseum Today

Even today, the Colosseum is exposed to earthquakes from Italy's earthquake source zones. Even today, there is a risk of damage to those parts that are not strengthened and treated. The Colosseum is also exposed to vibrations caused by the subway that passes beneath the Colosseum. Although the vibrations are quite intense, they are at relatively high frequencies that cannot excite the most important vibrational modes that would damage the Colosseum.

Is It Possible to Twist the Monumental Columns of Trajan and Marcus Aurelius in Rome?

Trajan's Column

On the occasion of the victories of Emperor Trajan (reigned 98–117 AD) in the Dacian Wars, a 35 m high column of Carrara marble was erected on the Forum Traiani. The monumental column was evidence of Trajan's extremely successful foreign and domestic policy. During his reign, the Roman Empire reached its greatest territorial extent.

The construction of the column was entrusted to Trajan's experienced architect Apollodoros of Damascus. Inside the column he created a spiral staircase, unusual for Rome at the time. The bas-relief depicted more scenes from the lives of ordinary Romans than scenes of battle. The emperor probably wanted to please the Roman people as a just and moderate ruler. The bas-relief could be admired from the two two-storey library buildings, concentrating the most important Greek and Latin intellectual works. From the windows of the upper floors it was thus possible to see 155 scenes made up of 2662 figures with the naked eye.

The top of the column was originally adorned with a statue of the emperor, and the emperor's ashes were placed in the base of the column. The statue was replaced by a statue of St. Peter in 1587 by order of Pope Sixtus V (1521–1590, pontificate 1585–1590).

Column of Marcus Aurelius

Trajan's column and its unique composition served as a model for other emperors who wanted to leave visible evidence of their successful military campaigns for future generations. In 180 AD, Emperor Commodus (reigned 177–192 AD) ordered the construction of such a column in honour of his father and previous emperor Marcus Aurelius (reigned 161–180 AD) (Fig. 8.3). The pillar was credited with the successful suppression of

Fig. 8.3 The famous column of Marcus Aurelius in Piazza Colonna in the historic centre of Rome, built according to Trajan's Column, located on Piazza Foro Traiano. Reprinted with permission from Moczo et al. (2023), © GRADA Slovakia s.r.o., 2023. All rights reserved

Marcomanni, Sarmatians and Quadi north of the Danube, also in the territory of today's Slovakia.

The wars lasted from 166 to 180 and the remains of the Roman presence in Slovakia can still be admired today: the inscription on the castle rock in Trenčín dates back to 179, when the Roman legions wintered in these parts. The emperor, also known as "the philosopher", thanks to his admirable education in philosophy, was to write the fundamental work of Stoic philosophy, *Meditations*, during the campaigns at river Hron.

The column was also known under the name Centenaria because it measured 100 Roman feet, i.e., 29.6 m. Like Trajan's statue, the statue of Marcus Aurelius on the top of the column was replaced in the sixteenth century by the statue of the Christian apostle St. Paul.

The frame of the column is essentially a copy of Trajan's column, but the relief is more dramatic and contrasting compared to that of Trajan's column. There was no similarly tall building near the column and the column could only be observed from below. Therefore, facial expressions are more expressive and less realistic. The scenes depicted the brutality of the war conflict somewhat more objectively.

Columns and Earthquakes

Both columns are 360 cm in diameter and are approximately 40 m high, including bases and statues. The column shaft consists of 17 blocks (hollow cylinders) of Carrara marble. The blocks were built on top of each other without mortar. The two blocks were connected together by four metal pins embedded in drilled holes and subsequently sealed in place with fused lead. However, metal pins and lead were looted in the Middle Ages, when monuments became an easy source of metal for swords and daggers.

Since the columns had the same material and construction structure, it would be understandable if both columns were in similar states today. However, this is not the case. The column of Marcus Aurelius was significantly more damaged in the past. Cracks have appeared and even heavy marble blocks have been dislocated! The column of Marcus Aurelius has therefore been repaired several times. Although the repairs have masked some of the damage, it is still possible to see a clear dislocation of the blocks by 8.2 cm due to the horizontal displacement and rotation of the two blocks relative to each other (Fig. 8.4).

What could have caused the mutual movement of two marble blocks, each weighing more than 30 tons? It's hard to imagine anything other than an

Fig. 8.4 Although the columns of Trajan and Marcus Aurelius were built of the same material and structure, the dislocation and twisting of heavy marble blocks can only be seen on the column of Marcus Aurelius. Trajan's column is undamaged. The explanation is the anomalous seismic motion in sediments under the column of Marcus Aurelius. Reprinted with permission from Moczo et al. (2023), © GRADA Slovakia s.r.o., 2023. All rights reserved

earthquake. However, there is a question. The pillars are less than 700 m apart. That is less than 1% of Rome's distance from the central Apennines, where earthquakes with potential macroseismic effects in Rome originate. Why did the earthquake damage only one of the two nearby pillars?

Strong Site Effect

The only possible answer is a significantly stronger local seismic motion beneath the column of Marcus Aurelius compared to the seismic motion of the bedrock on which Trajan's column stands. Only local conditions can cause such a significant difference in seismic motion. Trajan's column is situated on a consolidated layer of volcanic and Quaternary sediments underlain by solid clay rock. The column of Marcus Aurelius is situated on younger unconsolidated sediments, under which there is bedrock as under the Trajan's column. These sediments form a shallow valley bounded in the west-east direction. The position of both columns is shown in an image that is a modification of an image from an article published by Enzo Boschi et al. in 1995 in the *Bulletin of the Seismological Society of America*.

For seismologists, the configuration itself is already a strong indication that the seismic motion beneath the column of Marcus Aurelius may be significantly stronger. The qualitative estimate was supported by numerical simulations of the seismic motion. In this case, the seismic motion results from two wave phenomena. One is the so-called vertical resonance caused by the interference of waves multiply reflected between the free surface of sediments and the interface between the sediments and stiffer rock beneath the sediments. The other is the generation and propagation of surface seismic waves in a shallow sediment valley. Both processes contribute to the fact that the amplitude of the wave entering the valley is amplified by a factor of up to 7 at frequencies close to 1 Hz. The numerical-modelling results were also supported by measurements of ambient seismic noise at the locations of both columns.

Based on the analysis of earthquakes with an epicentre in central Italy and numerical modelling, it can be roughly estimated that the column of Marcus Aurelius could have been exposed to a horizontal acceleration of more than 1 g.

The Italian seismic catalogue lists several powerful earthquakes with an epicentre in the central Apennines, which had macroseismic effects in Rome. Seismologists from the Istituto Nazionale di Geofisica e Vulcanologia searched for a causal earthquake for damage to the column of Marcus Aurelius in the period between the theft of metal pins (thirteenth—fourteenth century) and the restoration of the column ordered by Pope Sixtus V. The first time determines when the structure of the column was weakened and more sensitive to earthquake motion. The second time determines the time of extensive repairs and reconstruction of the bas-relief. The current integrity of the bas-relief excludes any significant damage to the column after this reconstruction.

The only strong earthquake recorded in Rome during this period was the 9 September 1349 Apennine earthquake, which caused widespread damage, as documented by many authors, the most famous of whom is the Italian Renaissance poet Francesco Petrarca (1304–1374).

At present, without any alternative explanation, the most likely explanation for the damage to the column of Marcus Aurelius described above seems to be the mutual resonance in the sediments and the column during this earthquake. The measured natural frequencies of both columns are around 1 Hz.

In addition to representing an important part of the archaeological history of Rome, the two columns provide a warning for modern structures located on the sedimentary bedrock of the city. Especially for those buildings whose natural frequencies (the frequencies at which buildings willingly vibrate) are close to 1 Hz, which usually corresponds to up to 10-storey modern buildings.

9

Apocalyptic Tsunamis in Sumatra and Japan

The Tsunami Story Told by Friends

Phuket in Thailand is an island with long beaches and good facilities, hospitable to families with children, tourists, and retirees from cold countries who spend their vacation here. It is located approximately 420 km northeast of the city of Banda Aceh on Sumatra.

It was the year 2004, and my friends, Mirka and Peter on their way through Asia stopped in Phuket to spend a peaceful Christmas holiday there. They stayed at the Kamala Beach Hotel & Resort on the west coast. The Boxing Day was a bit busier at breakfast, but those who slept well didn't notice any tremors, and more sensitive types attributed them to lively holiday celebrations.

During breakfast, the creeping low tide caught the attention of guests, leaving all the boats stranded soon after. Some of the more curious guests went to see how far the water had receded and what lay so far out on the seabed. The water had withdrawn approximately 300 m. Such an unexpected ebb of water did not disturb the guests, and the locals were busy with checking their boats.

After a while, however, the water began to return. Peter recalls suddenly realizing that the birds had fallen silent. He saw an unusually high wall of water. Shortly after, he heard a German tourist shouting, 'Achtung, tsunami, achtung, tsunami!' Guests and locals alike panicked and fled up the stairs. No one had time for worries or thoughts. Fortunately, some sort of collective instinct of perceived danger kicked in. Everyone ran up the stairs and through the inner balconies as high as they could.

The wave reached the second floor and receded, taking cars, motorcycles, and anything that wasn't secured or anchored. It swept away people who weren't in a solid elevated place. Everywhere around, there was devastation and an eerie emptiness.

As a crisis manager, Mirka immediately realized the need to retrieve documents and money from the safe, pack a bag, which she placed on top of the wardrobe in case of another wave. She also grabbed bottles of drinking water and alcohol from the minibar. This turned out to be a wise decision, as the alcohol was used to treat the wounds of injured tourists and staff.

A few dozen hotel guests and staff managed to climb onto the roof in time. Some guests were in rooms on higher floors, and others climbed along the balconies outside to escape the rising water. The footage from the Kamala Beach Hotel & Resort later circulated the world.

Fortunately, the height of the tsunami at that location did not exceed 5–6 m. Although the hotel was built on a sandy ground, it withstood the tsunami. This was thanks to the 7 m concrete piles on which the hotel was built. The surrounding smaller hotels, apartments, makeshift homes, or restaurants did not have such luck. They disappeared.

People were already waiting on the hotel's roof for the second crest of the tsunami. Nobody knew what would happen next. Chaos ensued, and few dared to venture out. Electricity gradually failed, mobile networks went down, and desperation grew. People borrowed clothes and searched for relatives.

The second crest arrived within a few tens of minutes. It was like a black wall getting closer to you. That black wall kept growing in height, and you painfully realized your helplessness. You knew that the water would rise so high that it would engulf, suffocate, and sweep you away. However, the wave suddenly broke sharply on the shallow shore and surged into the hotel. This happened to this and dozens of other similar hotels along the coast.

The water receded again, and the sea calmed down. It was radiant, blue, and reflecting the sunlight. However, the originally built-up coastline was swept away, and the access roads filled with bewildered tourists and desperate locals. Bodies of people began to wash ashore only after several days.

Tsunami

The largest tsunamis in history are among the most tragic natural disasters. When triggered by earthquakes, they caused significantly greater damage in inhabited areas than the earthquakes themselves. This is related to where and how tsunamis originate.

My friends and other tourists survived by luck. Even on the hotel roof, they had no chance of knowing what would follow after the first crest of the tsunami. As we will mention later, the only reliable way is to evacuate in time—either based on early warning or recognizing the first signs of a possible tsunami—to a sufficiently high point as far away from the coast as possible.

How Does Tsunami Form

We will explain the origin of tsunamis using terms familiar from elementary school.

Kinetic and Potential Energy Imagine, for example, a small metal sphere suspended by a thin solid filament. The sphere oscillates. In its central position it has the greatest velocity and therefore the greatest kinetic energy. At the extremes, the speed of the sphere is zero and therefore the kinetic energy is zero (Fig. 9.1).

With the displacement it is precisely the opposite. In the middle position, the displacement is zero and the so-called potential energy is therefore zero. In the extreme positions the displacement is largest and therefore the potential energy is greatest there. As the sphere oscillates, the kinetic energy changes to

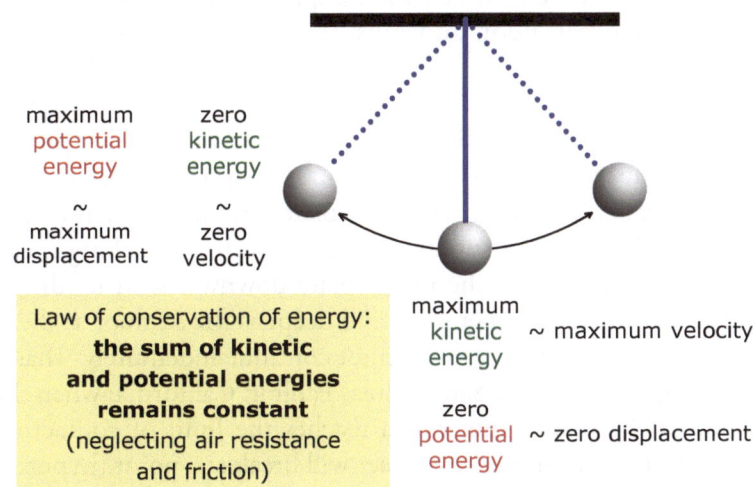

Fig. 9.1 The oscillating small sphere retains its total energy if air resistance and friction are neglected. Only the ratio of the kinetic and potential energy changes during the oscillation. This is important to realise for tsunamis. If the velocity of propagation, and hence the kinetic energy, is reduced, the potential energy, and hence the height of the tsunami, increases

potential energy and vice versa. However, the sum of the two energies is the same at each instant (if air resistance and friction are neglected).

Shallow and Deep Water If the bottom of a pond is 1 m deep, the water is shallow for an adult. If we're sailing on a boat or ship across the ocean and the ocean floor is 5 km deep, the water is deep for us. Both statements are undoubtedly correct.

If we throw a small pebble into the calm surface of a 1 m deep pond, a small circular wave will start spreading from where the pebble hits water surface. The distance between the crests of the wave (known as the wavelength) might be, for example, 3 cm. For such a short wave, a 1 m deep pond is very deep water. The ripple "doesn't feel" the deep bottom.

If a portion of the seabed or ocean floor rises abruptly during an earthquake, a massive amount of water is lifted, for example, by 1 m. The more and the faster the seabed rises, the higher the rise of the water mass. This water mass has tremendous potential energy in the Earth's gravitational field.

The lifted water mass immediately spills out into the surrounding area. Its potential energy quickly converts into the kinetic energy of the wave, which begins to propagate from the place of uplift. Depending on the specific situation, the wavelength (distance between wave crests) can be several tens of kilometres or even 500 km. Even a 5 km deep ocean is very shallow water for such enormous wavelengths. A wave with such a wavelength "feels" the ocean floor, and thus, the entire mass of water from the surface to the bottom.

Tsunami Generation in the Subduction Zone During the "preparation" phase for a subduction-zone earthquake, a lithospheric plate tries to slide under another plate, let's call it the upper plate. Friction at their contact (the subduction fault) causes the subducting plate not to slide along the contact surface, but to slowly "pull" the upper plate downwards in its direction of motion. This deforms the upper plate and deepens the seafloor at the contact of the plates. However, this process cannot continue indefinitely. That is similar to what happens along the San Andreas Fault in California: when the stress accumulated at the plate contact area reaches the limit of contact strength determined by static friction, the contact will break at a point (hypocentre)—a rupture will form. The rupture begins to propagate rapidly across the contact area of the plates. As the rupture propagates, the sinking plate is rapidly displaced in the direction in which it has a long-term tendency to move, and the tip of the upper plate, which has been pulled downwards before the rupture, suddenly "bounces" upwards. This causes a sudden uplift of the seafloor and a large mass of water (Fig. 9.2).

A Tsunami in the Open Ocean The wave that propagates from the uplift of the water mass due to the uplift of the ocean floor through the entire layer of water is called a tsunami. If we consider the height of the uplift of the ocean floor, for example, 1 m, the height of the wave crest (wave amplitude) above the place of uplift can be up to 1 m. It is understandable that due to the wave spreading in all directions, the height of the crest gradually decreases. The difference in height between the crest and the trough of the wave is therefore less than 2 m. Since the crest of the wave is tens or hundreds of kilometres away from the trough of the wave, it is clear that a tsunami goes unnoticed by humans and is "unfelt" by both small boats and large ships in the open ocean.

However, what is remarkable is the velocity at which a tsunami propagates: 600 to 900 km/h. The speed of a jet aircraft! This velocity increases with the depth of the ocean floor, i.e., with the thickness of the water layer. However,

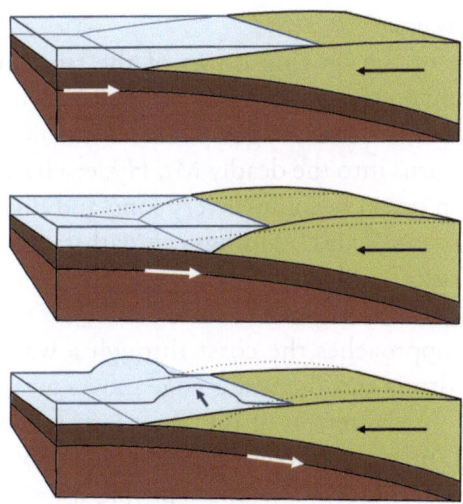

Fig. 9.2 In subduction zones, one lithospheric plate tries to slide under the other plate. Due to friction, it "pulls" and hence deforms it. After the accumulation of critical stress, a rupture arises and propagates over the plate contact. The result is a sudden displacement of the dipping plate in the dipping direction and a sudden uplift of the deformed edge of the upper plate. This uplift causes an uplift of the ocean water mass. Consequently, a tsunami is generated. Reprinted with permission from Moczo et al. (2023), © GRADA Slovakia s.r.o., 2023. All rights reserved

this means that even though a tsunami in the open ocean is absolutely harmless, as it has a small amplitude, it possesses truly enormous kinetic energy since the kinetic energy is proportional to the square of the wave propagation velocity.

Tsunami on the Coast In shallow water, the velocity of the wave decreases as the thickness of the water layer decreases, i.e. as the depth of the bottom decreases. As we have already said, any ocean is a shallow water for a tsunami. Therefore, as the tsunami approaches the coast, it "senses" the decreasing depth of the seabed and the velocity of its propagation decreases. As the velocity of the tsunami decreases, so does its kinetic energy. If we neglect friction losses, the total energy of the tsunami is conserved according to the law of conservation of mechanical energy. Therefore, as the kinetic energy decreases, the potential energy must increase (as in the case of a pendulum). However, this means that the amplitude of the wave increases, i.e. the wave crest height increases.

The distance between two wave crests (wavelength) equals the product of the wave's velocity and its period. Since the period of the wave remains constant, if the velocity decreases, the wavelength must also decrease. As the wave amplitude increases and the wavelength decreases, this results in an increase of the steepness of the wave.

As a consequence of the slowing down of the tsunami propagation, the velocity of the tsunami on the coast is only 50–300 km/h, and the wavelength is only 1–3 km. However, the height of the wave crest can reach up to 30–40 m. Dr. Jekyll turns into the deadly Mr. Hyde: a harmless and unnoticeable wave in the open ocean becomes a very long and high wall of water that only the sturdiest structures and buildings can withstand. Therefore, the largest tsunamis cause greater damage in coastal areas than the earthquakes themselves.

When a tsunami approaches the coast through a wave trough, the water level near the shore drops and the water edge moves away from the shore (it can be up to 1 km!). As the retreat of the water lasts for several minutes (half the tsunami period), many unknowledgeable people go out to collect shells and sea animals left on the bottom. These people usually have no chance of surviving the arrival of the crest of the big tsunami.

As the tsunami approaches the coast with a wave crest, the water level near the coast rises and the lower areas are flooded. After the water recedes, unknowledgeable people do not realise that another wave crest, usually higher than the first, follows.

Where the Tsunamis Occur

Tsunamis most commonly occur where subduction zones exist, where underwater earthquakes occur, and where is uplift of the seabed. However, tsunamis can also be caused, for example, by massive underwater landslides. And an underwater landslide can be triggered by earthquakes outside subduction zones. Another possible cause may be a massive collapse of a volcano's slope into the sea. In the case of underwater subduction earthquakes, the key factor for tsunami generation is the size of the earthquake, i.e., its moment magnitude.

Recall the tsunami that completed destruction begun by the earthquake of 1 November 1755 in Lisbon. Tsunamis are also generated in the Mediterranean. And they can be like the 2004 tsunami off Sumatra or the 2011 tsunami off Japan. Hawaii, the Canary Islands and many other holiday destinations are also places tsunamis can hit.

Tsunami Warning Systems Save Human Lives

Tsunami warning systems depend on a network of seismic stations and a sea level monitoring system. Early detection of a submarine earthquake in a subduction zone and a sufficiently accurate estimate of its magnitude are essential.

Warning is made possible by the fact that seismic P-waves travel much faster than tsunamis themselves. Because the P-waves arrive at seismic stations fast enough, it is possible to use a computer program to determine where the earthquake originated and how large it is. Seismologists are constantly refining these location programs and estimating the size of an earthquake due to advances in earthquake monitoring, computational algorithms and computers.

However, for issuing a warning, the so-called earthquake mechanism—the orientation of the fault surface (plate contact) where the rupture occurred and the propagation of the rupture on the fault surface—needs to be calculated quickly. This is a challenging mathematical and physical problem. Methods for solving it are still being developed.

The National Oceanic and Atmospheric Administration (NOAA) in the USA operates two tsunami warning centres. The Pacific Tsunami Warning Center (PTWC) was established as a result of internationally coordinated efforts following the experience of the 1960 Chilean earthquake-triggered tsunami. The centre is located on the small island of Ford in Pearl Harbor Bay on the Hawaiian island of Oahu. This centre covers the Pacific Ocean and the

Caribbean region. The National Tsunami Warning Center (NTWC) was founded in 1967 after the Great Alaska earthquake in 1964 and has its headquarters in Palmer, Alaska. It covers the continental United States, Alaska, and Canada.

The USA officially began issuing tsunami warnings in 1949, after a 14 m high tsunami, triggered by an Mw 8.6 earthquake near the Aleutian Islands, killed 160 people and caused massive damage in the city of Hilo on the big island of Hawaii.

The tragic tsunami on 26 December 2004 in Sumatra led to development of the German-Indonesian Tsunami Early Warning System (GITEWS).

However, there are areas where strong tsunamis can occur, but warning systems are either absent or not at the level of NOAA centres. Especially in these areas, having at least basic knowledge about tsunamis can be crucial for the survival of both tourists and locals.

Issuing reliable and accurate tsunami warnings requires understanding the potential tsunamis that could occur in a given coastal area. Since tsunamis most commonly result from underwater earthquakes, it is essential to know or estimate what kind of tsunami could occur if an earthquake of a certain magnitude were to happen in a specific location within a subduction zone. This can only be achieved through complex numerical simulations of earthquakes and tsunamis. Such simulations require sophisticated numerical methods and high-performance parallel supercomputers.

What Is Good to Know in a Place That Could Be Hit by a Tsunami

Minimal knowledge preparation is necessary, especially when choosing a place for a seaside vacation. At a certain height of the tsunami, it is really a problem to survive. For example, the NOAA website can be used to find out if a potential vacation spot may be affected by a tsunami. If so, you need to remember what a tsunami is and how you should act before and during a tsunami. At the vacation spot, finding escape routes from the beach to the nearest hill or solid tall building is essential.

If you do not feel tremors, but you observe a relatively sudden change in sea level and hear the roaring sound of the sea, it may indicate a distant tsunami, which could reach the coastline where you are in a few hours. If you feel the Earth tremors and hear the roaring sound of the sea, it may indicate a nearby earthquake, and the tsunami could arrive in a few minutes. If you notice water receding from the shore, the crest of the tsunami could arrive in a few minutes. If you see an unusual wave approaching the coast, it is likely a tsunami.

If you recognize signs of a possible tsunami, do not wait for official warnings and quickly move away from the coastline, preferably to a solid place at a sufficient height above sea level. If you already see a tsunami, evacuate to higher ground as quickly as possible.

Volcanoes, Earthquakes, Tsunamis and Megatsunamis

Below, we will pay particular attention to two devastating tsunamis that occurred in 2004 near Sumatra and in 2011 near Honshu. But before that, we will briefly mention the stories of five other important tsunami events from history.

Santorini 1650–1600 BC

Some readers will recall that one of the few natural disasters mentioned in school history lessons was the explosive eruption of the volcano on the Thera archipelago, now known as Santorini. The eruption ejected some 30–40 km^3 of material, caused the collapse of the volcanic island and triggered an unusually high tsunami. This event is thought to be responsible for the collapse of the Minoan civilisation on the island of Crete, about 110 km from Santorini.

In reality, however, the story is not so simple. The Minoan civilisation was dramatically affected by the earthquake, volcanic eruption and tsunami, but it is not clear to what extent. In the past, it was thought that volcanic dust covered much of Crete's surface, killing vegetation and causing famine. However, subsequent research has shown that the amount of ash that fell on Crete could not have caused such a mass die-off. Archaeological research suggests that the navy, which was crucial to the survival of the island's Minoan civilisation, and many of the coastal settlements may have been destroyed by the tsunami or by an earthquake and subsequent fire.

According to estimates, the eruption had a magnitude of 6–7 on the VEI scale. The VEI scale (Volcanic Explosivity Index) has 8 degrees and classifies volcanic eruptions according to several factors, namely the amount of pyroclastic material produced, the height of the ash cloud, the duration of the eruption's main phase and the volcano's overall activity. However, it also takes qualitative descriptive parameters into account. Grade 8 indicates eruptions that eject more than 1000 km^3 of material, grade 7 those that eject approximately 10 times less. The Santorini eruption may have been larger than the Krakatoa eruption (VEI-6) and may be comparable to the 1815 Tambora eruption (VEI-7).

The eruption of the Tambora volcano caused a dramatic decrease in temperatures, which in Europe resulted in the so-called The Year Without Summer in 1816, when temperatures dropped by 0.4–0.7 °C on average. All over the world, the temperature drop caused various degrees of crop failure and subsequently famine, epidemics and tens of thousands of victims. The so-called volcanic winter probably also occurred as a result of the eruption on Santorini.

Let's just remind that approximately 74,000 years ago the Toba supervolcano in Sumatra, Indonesia, erupted with a magnitude of VEI-8. This eruption is believed to have been so destructive that it almost caused the extinction of mankind.

Krakatoa 1883

Indonesia has 127 active volcanoes, more than any other country in the world. The west coast of Sumatra and the south coast of Java are literally littered with active volcanoes, which in the past have manifested themselves in huge eruptions. This is because a subduction zone runs along these coasts, where the Indian plate subducts under the Sunda plate. Above this subduction zone in the Sunda Strait is the archipelago of Krakatoa or Krakatau.

In May 1883, craters on the archipelago began spewing ash. The increased volcanic activity culminated in four eruptions on 27 August 1883. The third of them was accompanied by the sound of an explosion considered to be the loudest sound ever. The sound of the explosion was reportedly heard over 10% of the planet's surface. On the island of Mauritius, the sound was mistaken for the sound of cannons, in Batavia the sound reached up to 180 decibels (as if the observer was standing next to a launching space rocket), and anyone within 20 km of the crater would lose their hearing permanently, as the explosion there could have been calculated to be 310 decibels.

The eruption of the Krakatoa volcano (VEI-6) ejected 25 km^3 of material and ash and was compared to the explosion of 200 megatons of TNT. The impact of a huge volume of material into the ocean caused a tsunami. The tsunami with a height of up to 30–40 m was supposed to be responsible for the largest number of victims out of a total of 36,000.

Meiji Sanriku 1896 and Kantō 1923

The question of whether there is a connection between earthquakes and tsunamis was raised by Japanese seismologists following the earthquake that struck the Sanriku region, North-Eastern Honshu, in 1896. The earthquake, estimated to have a magnitude of 8.5, occurred on 15 June at 7:32 PM local time. Thirty-five minutes later, a tsunami, with then the largest observed height of 38 m, struck unsuspecting people, claiming 22,000 lives.

The Japanese seismologist Akitsune Imamura (1870–1948) perceived the catastrophe very personally. In the years 1897–1900, he developed a theory on the origin of tsunamis—the movement on underwater faults during earthquakes causes the uplift of a massive amount of water, generating waves that peacefully propagate across the ocean. The problem for people arises when the waves approach the land and transform into a steep and high water wall, sweeping away everything in its path.

It is worth mentioning that Akitsune Imamura also made history with his consideration of a future major earthquake. In his work in 1905, he stated that the Kantō region on the eastern coast of Honshu would be struck by an earthquake in the following half-century, resulting in more than 100,000 casualties.

In fact, on 1 September 1923, the so-called Great Kantō Earthquake occurred with an estimated moment magnitude of 7.9–8.2. It devastated Tokyo, Yokohama, and the prefectures of Chiba, Kanagawa, and Shizuoka. It caused a fire and triggered a 10–12 m high tsunami. The fire erupted into a firestorm that spread uncontrollably. As a result of all the elements, more than 100,000 people died.

There is a lesson for society in every disaster. The 1896 earthquake and tsunami gave rise to one of the first scientific explanations of the relationship between earthquakes and tsunamis. Since 1960, the 1 September has been Japan's Disaster Prevention Day, commemorating the importance of natural disaster preparedness and disaster preparedness drills in schools and other institutions. Akitsune Imamura emphasised the importance of education and preparation in an attempt to reduce the number of victims.

Until 2011, the tsunami triggered by the earthquake in 1896 was considered the most devastating in documented history in Japan. It was on the occasion of the tsunami that struck the Sanriku area that the term "tsunami" (tsu = harbor + nami = waves) was first used in the English language, derived from the Japanese 津波.

Messina 1908

The danger posed by the Messina Strait separating Sicily from southern Italian Calabria has been recognized by its inhabitants and visitors for millennia.

In Greek mythology, the Messina Strait was referred to as Scylla and Charybdis. Charybdis, the daughter of Poseidon, the god of the sea, and Gaia, the goddess of the Earth, took the form of a whirlpool, greedily sucking everything in its path into the depths of the sea. Across the strait in a cave lay the voracious Scylla with 6 heads and 12 feet. This ancient Greek myth encapsulated the peril that the Messina Strait presented to sailors.

The local folk tale of Cola Pesce, half boy and half fish, was one of the ways people explained the frequent earthquakes for centuries. According to the legend, the king of Messina ordered the boy to find out what lay at the bottom of the sea below Messina. Cola Pesce dived below the surface several times. Then he said to the king Messina was built on three pillars, one of which stood firm, another was broken and the third had disappeared completely. This legend raised the question of what would happen to Messina if the last stable pillar gave way.

On 28 December 1908, although the column did not break, and the angry Scylla and Charybdis did not awaken, there was certainly a dynamic and dramatic movement at a depth of about 10 km near Messina. A Mw 7.1 earthquake occurred. It is not clear whether due to the earthquake or an underwater landslide triggered by the earthquake, but, shortly after the earthquake a tsunami occurred. Tsunami hit the coast of Sicily and Calabria in particular, reaching a height of up to 12 m. The towns of Messina and Reggio Calabria, just 10 km away, were almost completely devastated. An estimated 60,000 to 100,000 people died as a result of the earthquake and tsunami, but it is not clear how many of these deaths were caused by the tsunami itself.

The cities descended into chaos; surviving prisoners escaped from the destroyed prisons, exacerbating the already low morale. Fires spread, and the telegraph network became non-functional. Only fragmentary reports reached Rome and were not given proper attention. The government and army mobilized only after they were alarmed by Russian and British ships sailing through the Mediterranean, which were the first to reach the devastated Messina. The delayed mobilization was caused by the unstable political situation and Rome's indifference towards southern Italy, perceived as backward and provincial, an area of uneducated workers from citrus plantations, despite its contribution to Italy's lucrative export industry.

Alaska 1958—Megatsunami

The Fairweather Fault, also known as the Queen Charlotte Fault after the nearby Queen Charlotte Islands, is a right-lateral transform fault similar to the San Andreas Fault in California. It marks the boundary between the North American and Pacific Plates, running along the southern coast of Alaska.

On 9 July 1958, at 10:15 PM local time, an earthquake with a magnitude of Mw 7.8–8.3 occurred on the Fairweather Fault. The earthquake triggered a massive landslide in the Lituya Bay (Fig. 9.3), located 60 km away, where 90 million tons of rock and soil collapsed into the water from the steeply rising coastline. The noise accompanying the landslide could be heard up to 80 km from the bay. The landslide material entering the bay triggered a so-called megatsunami, which reached the highest recorded height—an astonishing 524 m. The wave swept away 4 km² of forest cover from the surrounding area, leaving behind only bare rocks, soil, and remnants of vegetation.

The name megatsunami may give the impression that this is an extremely large tsunami. As we explained at the beginning of this chapter, tsunamis that can be described as common or normal are caused by the uplift of the seabed

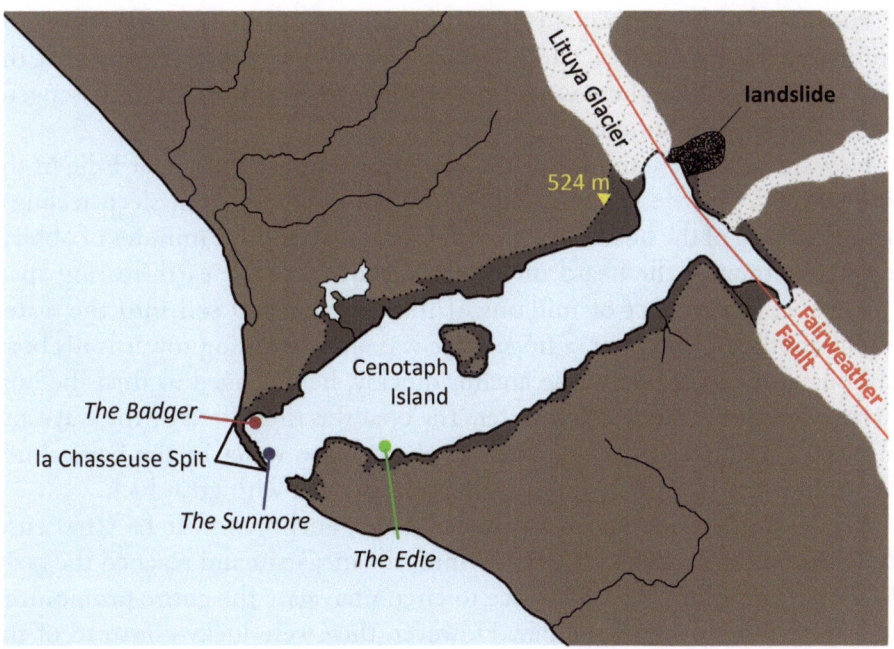

Fig. 9.3 Lituya Bay in Alaska, where a megatsunami was caused by an induced landslide. Reprinted with permission from Moczo et al. (2023), © GRADA Slovakia s.r.o., 2023. All rights reserved

or by an underwater landslide. These tsunamis gain great height in shallow water as they approach the coast. The term megatsunami is used for an extremely high wave caused by, for example, a sudden landslide of a huge amount of material into the water. Such as in Lituya Bay, or a landslide of part of a volcanic hill into the sea.

The height of the tsunami in Lituya Bay was aided by the geometry of the bay. The bay is essentially a fjord in the shape of a "T", with the Lituya and Crillon glaciers located on its two arms. The land rising from the sea is steep. It is a continuation of the bottom, which is deepened into a "U" shape within the bay. In the middle of the bay lies a small island called Cenotaph. The mouth of the bay is narrowed by a kilometre-long promontory called La Chasseuse. The bay has a length of 14.5 km and a width of 3.2 km.

The earthquake and megatsunami in 1958 sparked interest among scientists in Lituya Bay. They found that the megatsunami in the bay was not an isolated incident. Significantly high tsunamis likely occurred here also in 1854, 1899, and 1936. However, already Jean-François de Galaup La Pérouse, a French explorer, described a peculiar sharp line of forest growth on the rocks surrounding the bay. While mapping the bay in 1786, he noticed it was located well above the waterline.

Fortunately, the earthquake and megatsunami primarily affected nearly uninhabited areas of Alaska and resulted in only 5 casualties in total. At the time of the landslide, there were 3 boats in the bay, each with two passengers: Edrie, Badger, and Sunmore.

Howard G. Ulrich and his 7-year-old son anchored their boat Edie in the southeast part of the bay. Howard G. Ulrich woke up from sleep feeling a strong rocking of the boat. He went on deck and after three minutes of observing strong tremors, he heard the deafening sound of the earth tearing apart and the violent impact of millions of tons of rock and soil into the water. About another 2 min later, a huge wave was already rolling towards his boat. Although he couldn't raise the anchor quickly, he managed to turn the boat facing the approaching wall of water. The boat was thus lifted by the wave and survived about 30 min of intense waves until the water in the bay calmed down. Howard G. Ulrich and his son thus survived with great luck.

Mr. Swanson woke up on the boat Badger, anchored near La Chasseuse, due to strong tremors. After about 3 min, the megatsunami reached the boat. The Swansons didn't have a chance to circumnavigate the entire promontory and reach the mouth of the bay. However, they were lucky—instead of the wave sweeping their boat away, the wave lifted it, tearing the anchor from the bottom, and carried the boat over the promontory to the open sea. The wave destroyed the boat, but the Swanson couple managed to get into a small rescue dinghy and survive.

Paradoxically, the Wagner couple, who were anchored with the boat Sunmore closest to the mouth of the bay, were unable to save themselves.

Three more people perished due to the earthquake, which also caused a landslide of part of Khantaak Island, 160 km north-west from Lituya Bay.

2004 Sumatra–Andaman Earthquake and Tsunami

Tectonic Situation

Earthquakes that can generate tsunamis off Sumatra originate in the subduction zone where the Indian Plate dips (subducts) beneath the Sunda Plate. The contact between the Indian and Sunda plates is formed by a large fault (interplate thrust or megathrust). Subduction has resulted in the formation of the Sunda (formerly Java) Trench, which is approximately 3200 km long and has a maximum depth of 7290 m. The arc of the trench extends from the Lesser Sunda Islands south around the island of Java and southwest around the island of Sumatra to the Andaman Islands. The bottom of the trench defines the line where the Indian Plate meets the Sunda Plate on the ocean floor. This means the line where the fault crosses the ocean floor. The subduction of the Indian Plate under the Sunda Plate is not simple. The Indian Plate moves relative to the Sunda Plate in a north-northeastern direction at a rate of 5.9 cm/year near the Andaman Islands and 6.8 cm/year in southern Sumatra. The dip angle gradually increases from 5–7° just off the trench floor to about 30° off the coast of Sumatra. The tectonic situation is complicated by transform faults within the Sunda Plate (notably the Great Sumatran Fault), which run parallel to the line of the Sunda Trench from the northern half of Sumatra to the Andaman Islands.

As a result of the subduction of the Indian Plate under the Sunda plate, there are the already mentioned 127 active volcanoes in Indonesia. Up to five million Indonesians live in immediate danger from volcanoes.

Earthquake

On 26 December 2004, at 7:59 AM local time, a submarine earthquake occurred with a moment magnitude of Mw 9.1 (Fig. 9.4). One of the most prominent seismologists, Professor Hiroo Kanamori of the California Institute

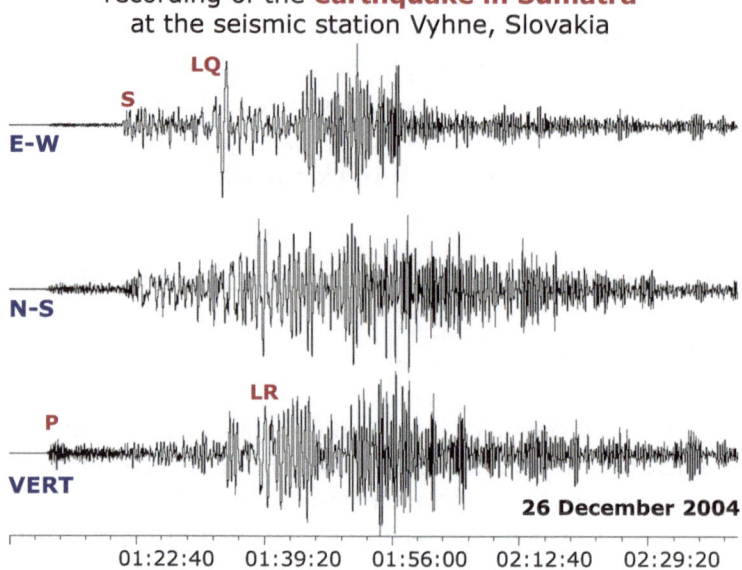

Fig. 9.4   The earthquake in Sumatra was recorded by seismic stations worldwide. The epicentre of the earthquake was approximately 8800 km away from the Vyhne seismic station in central Slovakia. VERT, N-S, and E-W denote the vertical, north-south, and east-west components of seismic motion. P represents the P-wave, S represents the S-wave, LR represents the Rayleigh wave, and LQ represents the Love wave

of Technology (Caltech), estimated the magnitude at Mw 9.2. Another independent analysis allows for Mw 9.3. This variability indicates that determining the size of the largest subduction earthquakes is not simple at all. If it were Mw 9.1, it would be the third largest seismologically recorded earthquake since 1900. If Mw 9.3, it would be the second largest earthquake after the Great Chilean earthquake in 1960. It is worth noting that Mw 9.3 releases approximately twice as much energy as an Mw 9.1 earthquake. The total seismic energy released by the earthquake was estimated at 1.4×10^{17} Joules. The energy released by the atomic bomb explosion in Hiroshima was 6.3×10^{13} Joules. This means approximately 2200 times less. The methods for estimating magnitude and radiated energy are different, and therefore it is not straightforward to directly correlate the respective estimates.

According to the United States Geological Survey (USGS), the hypocentre of the earthquake was at a depth of approximately 30 km. The epicentre was approximately 160 km west of the coast of Sumatra Island and 50 km north of the small island of Simeulue.

The rupture propagated from the hypocentre along a plate contact dipping at approximately 10° in a roughly north-northwest direction. It propagated relatively slowly with little slip for the first 40–60 s, then expanded at about 2.5 km/s, reaching velocities of up to 3 km/s. In total, the rupture extended over an area 1200–1300 km long. The slip reached up to 15 m over the approximately 600 km long segment.

Interestingly, and very importantly, the rupture in this earthquake propagated dominantly in one direction from the hypocentre. This dominant unilateral (or unidirectional) propagation was also found in other large subduction earthquakes—Mw 9.5 Chile 1960, Mw 9.2 Alaska 1964, and Mw 9.0 Kamchatka 1952.

Using GPS stations, static displacements of the Earth's surface up to 4500 km from the epicentre have been detected.

In addition to Indonesia, the earthquake was felt in Malaysia, Singapore, Thailand, Myanmar, Bangladesh and the Maldives.

The earthquake also triggered a large number of modes of Earth's free oscillations. These oscillations had anomalously large amplitudes at periods greater than 1000 s and were measurable for several weeks after the earthquake. Seismic records of the oscillations provided invaluable data, the analysis of which allowed for further insight into the interior of our planet. Similarly important data based on the free oscillations were obtained only during the Mw > 9.0 earthquakes in Chile 1960, Alaska 1964, Sumatra 2004, and Tōhoku 2011. The longest observable period due to this earthquake, 53.7 min, was for the so-called "football mode." It is worth noting that the name comes from American football—in this mode, the Earth deforms into the shape of a ball used in this sport (see Fig. 2.13).

Irreversible displacements on the ruptured part of the fault and in the surrounding area have caused changes in the distribution of mass inside the Earth. As a result, the North Pole has shifted by 2.5 cm and the flattening of the Earth has been slightly reduced. The Earth's rotation has accelerated slightly and the day has been shortened by 2.68 microseconds. Such small changes occur in earthquakes of this size.

The earthquake was followed by a series of aftershocks on and close to the ruptured fault. The largest aftershocks had a magnitude Mw 6.6. It is unclear whether the Mw 8.6 Nias-Simeulue earthquake of 28 March 2005 was an aftershock. Unclear because the maximum magnitude of aftershocks is usually at least one unit smaller than that of the main earthquake. Some seismologists believe this was an earthquake caused by a change in the distribution of stress and deformation due to the main earthquake, rather than an aftershock of the December earthquake.

Several experts suggest the earthquake may have influenced the preparation of eruptions of some volcanoes. Most importantly, it immediately generated a huge and tragic tsunami.

Tsunami

A few metres of uplift of the seabed over a length of several hundred kilometres along the fault line caused by an earthquake lifted a huge mass of ocean water. The subsequent gravitational descent of the uplifted water mass caused the tsunami. Because Indonesia did not have a warning system at the time (and should have had one long ago), the first warning and information about the possibility of a major tsunami did not come from the Pacific Tsunami Warning Centre until 15 min after the earthquake. Due to this and also due to the overall tragically low level of knowledge and awareness of tsunamis among locals and tourists, many people were needlessly killed when the tsunami hit the coast of the Indonesian province of Aceh within 20–30 min of the earthquake.

One of the positive exceptions is the evacuation of several communities in Sri Lanka thanks to individual warnings from relatives living abroad who learned about the impending tsunami earlier than locals.

In the province of Aceh, the tsunami reached a maximum height of at least 30 m, with the National Oceanic and Atmospheric Administration (NOAA) reporting up to 51 m. Such extreme values are usually very difficult to prove because such high tsunamis can be hardly survived by people or technical devices. Moreover, there may not be visible traces in nature from which the height of the tsunami could be determined.

The tsunami penetrated relatively deep into the interior and ruthlessly killed and destroyed. In the province of Aceh, tsunami penetrated up to 5 km from the coast. According to realistic estimates, nearly 170,000 people lost their lives in Indonesia, although in the catastrophic devastation of the country and in the settlement patterns, it was simply not possible to officially confirm tens of thousands of victims.

An hour and a half later, the tsunami reached the beaches of southern Thailand. Of the more than 5000 dead, 2000 were foreign tourists. Half an hour later, the tsunami hit the coast of Sri Lanka, killing more than 35,000 people. Shortly after Sri Lanka, the tsunami hit the east coast of India, particularly the area south of the city of Chennai. In some places, the tsunami reached a height of 5 m.

It is important to note that the west coast of Sri Lanka and India was also affected. Figuratively speaking, the tsunami also reached places that were relatively far "around the corner" or in the deep "shadow" of Sri Lanka and southern India.

On the western coast of Sri Lanka, the tsunami swept away a train traveling from the capital Colombo to the southern city of Galle. It was one of the largest railway tragedies ever, resulting in the loss of 1500 passengers aboard the train.

For example, in the area of Cochin, located approximately 270 km northwest along the western coast from the southern tip of India, the tsunami penetrated nearly 200 m inland. The maximum height of the tsunami reached 6 m there. In other words, the entire Indian subcontinent was no obstacle for the tsunami to reach practically the entire western coast of India.

The penetration of tsunamis onto the western coasts of Sri Lanka and India is a consequence of diffraction. Tsunami is a wave. All types of waves diffract at obstructions such as tips, edges or slits. A simple illustration is shown in the picture (Fig. 9.5).

Bangladesh has a low elevation, making it vulnerable to significant damage from tsunamis. However, a fortunate circumstance for Bangladesh was the orientation of the ruptured fault, which resulted in a relatively small tsunami height directed towards Bangladesh.

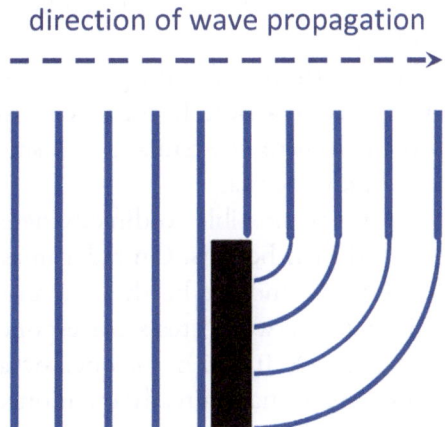

direction of wave propagation

Fig. 9.5 The picture illustrates wave diffraction. If a wave (for simplicity with a plane wavefront) encounters an obstacle, here represented by the black rectangle, it can, as a result of diffraction, penetrate beyond the obstacle at the upper right tip of the obstacle, as indicated by the quarter circles. Due to diffraction, the tsunami from Sumatra in 2004 penetrated, for example, to the western coasts of Sri Lanka and India

In its unstoppable expansion in the Indian Ocean, the tsunami also hit the touristically very attractive Maldives. According to estimates, the tsunami height did not exceed 4 m, in several places 2–3 m. The tsunami cannot typically increase in its approach to the Maldives because of the complex geometry of the ocean floor surface and because of the coral reefs. Despite this, approximately 100 people have died in the Maldives.

About 7 to 8 h after the earthquake occurred, the tsunami reached Madagascar and the easternmost coast of Africa. More than 300 people died in Somalia, Tanzania and Kenya. In Somalia, the height of the tsunami reached more than 9 m.

If we consider the distance between the origin of the tsunami and Madagascar to be approximately 5300 km, and the propagation time to be 8 h, we see that the tsunami propagated between the origin and the north of Madagascar at an average velocity exceeding 660 km/h.

The tsunami visibly affected 15 countries: Indonesia, Sri Lanka, India, Thailand, Somalia, Myanmar, Maldives, Malaysia, Tanzania, Seychelles, Bangladesh, South Africa, Yemen, Kenya, and Madagascar. In many places, it penetrated up to 2 km inland. It is estimated that a total of 230,000 people lost their lives in 14 countries.

Propagating between Africa and Antarctica, the tsunami also reached the Atlantic Ocean, hitting Greenland after about 30 h. It also reached the Pacific Ocean by propagating between Australia and Antarctica, and reached Alaska after approximately 32 h. The height of the tsunami may have been increased in some places due to local conditions. In Manzanillo, Mexico, it is said to have reached a height of 2.6 m.

The orientation of the fault rupture and the geometry of the tsunami propagation led to an interesting comparison. In the Cocos (Keeling) Islands, only about 1700 km from the epicentre, the tsunami was smaller than in Peru or the Canadian province of Nova Scotia.

For the first time ever, it was possible to directly determine the distances between tsunami crests, and their heights. On radar images from the Jason-1 satellite, launched into orbit around the Earth at an altitude of 1336 km by NASA in 2001, the distances between crests corresponding to the tsunami wavelength ranged from 500 to 850 km. In the open ocean, the tsunami only reached a height of 0.5 m. As we have already mentioned, with such a small amplitude and such a large wavelength, the tsunami is invisible and absolutely harmless in the open ocean. Even at the speed of a jet aircraft.

The total energy of the tsunami was estimated at 2×10^{15} J, which is said to be more than twice the total energy of the explosives used throughout World War II (including the two atomic bombs).

Approximately 1.7 million people became homeless. In 2017, the total damage was estimated at approximately 13 billion US dollars. The worst affected country was Indonesia, where damages amounted to nearly 6 billion US dollars.

The earthquake and tsunami together altered the landscape of many coastal areas of the Indian Ocean. Erosion and coastal subsidence caused some parts of the coastline to vanish into the ocean, while in other areas, coral reefs emerged above the surface.

The tsunami devastatingly affected coastal ecosystems—mangroves, marsh-lands, coral reefs and forests. The large amounts of seawater have degraded the soil and the micro-organisms therein. Groundwater supplies have been affected too. The water was also polluted by solid and liquid waste, which meant, among other things, a shortage of drinking water. This changed the natural habitat for many plants and animals, and made many areas inhospi-table to their natural inhabitants. Agriculture and aquaculture in Sumatra, Sri Lanka and other areas have dealt with the ecological consequences of the tsunami for decades. This aspect has also been part of foreign aid programmes.

The earthquake and tsunami caused enormous local economic burdens, particularly in the areas of agriculture, infrastructure, fishing, and tourism. While tourism and fishing contribute only a small portion to the GDP, they locally guarantee thousands of jobs, which were significantly reduced due to the tsunami. Consequently, they markedly increased the unemployment rate. The damaged agriculture and infrastructure had an impact on the national economy.

Tsunami: A Catalyst for Resolving the Political Situation?

Using the example of Sri Lanka and Sumatra, it is possible to demonstrate how a country can cope with problems resulting from a natural disaster and to what extent a natural disaster can influence seemingly unrelated political situation in the affected country.

Both countries were dealing with long-standing internal conflicts between the official government and separatist groups at the time of the earthquake and tsunami. However, the nature and outcome of these conflicts in the con-text of a natural disaster were different.

Aceh had been an independent Muslim sultanate since 1496. During the centuries of European colonization of Indonesia, the Sultanate maintained relative autonomy and contributed significantly to the colonial trade in black pepper. In 1950, it was incorporated into the Indonesian province of North

Sumatra, and the 1950s were marked by struggles for autonomy in accordance with Islamic law. These demands were partially met in 1959. However, in the early 1970s, it was discovered that Aceh was rich in oil and natural gas.

The province could have significantly benefited from its natural resources. However, the Indonesian government decided to confiscate private land to build an oil refinery. This act was the last drop for Aceh in a sea of ignored and suppressed efforts to gain autonomy and full religious freedom. It led to the formation of the separatist group, the Free Aceh Movement, in 1976. The catastrophic consequences of the tsunami led to the ceasefire of hostilities just 2 days after the tsunami.

The story of Sri Lanka is different. Since around the second century BC, the northern part of Sri Lanka has been inhabited by Tamil ethnic groups, who has also resided in the South Indian state of Tamil Nadu. In the nineteenth century, until gaining independence from Britain in 1948, the Tamil population increased. Tamil labourers were imported to Sri Lanka, then one of Britain's colonies, to work on plantations. The vast majority of Tamils profess Hinduism, while the majority (currently nearly 75%) of the native Sinhalese population profess Buddhism. Conflicts between the two groups of people of different ethnicities, languages, and religions intensified in 1956 when Sinhalese was recognized as the sole official language of Sri Lanka. The law seemed to trigger widespread oppression of the Tamil minority. In 1976 (the same year as in Sumatra), a militant nationalist group called the Liberation Tigers of Tamil Eelam (Eelam being the original Tamil name for Sri Lanka) emerged in Sri Lanka. Their goal was to establish an independent Tamil state on the island, encompassing the entire eastern, northern, and half of the western coast where the Tamil ethnic group was most populous. Since its inception, the Tamil Tigers engaged in aggression both against the official government and civilians. A ceasefire was brokered between the government and the separatists in 2002.

However, everything changed in 2004 when the tsunami devastated predominantly Tamil areas. After the disaster, another agreement was reached between both sides, aiming for a fair non-governmental redistribution of foreign humanitarian aid. Sinhalese nationalists disagreed with this, as they labelled the Tamil Tigers as a terrorist organization. Thus, ethnic politics also interfered with efforts to revitalize the island after the tsunami.

Until May 2009, when the Tamil Tigers were definitively defeated, the internal conflict intensified further. In the first 5 months of 2009 alone, more people died as a result of the conflict than from the tsunami.

The Tamil Tigers are considered one of the most materially and organizationally secured terrorist groups in the world. They had their own navy, air force, suicide unit, and controlled territory with subordinate administration and tax collection, which financed their activities. They also derived finances from piracy, although this activity significantly declined in the entire affected area after the tsunami because most ships were destroyed.

The conflicts on Sumatra and Sri Lanka claimed tens of thousands of lives, with about five times more casualties on Sri Lanka than on Sumatra. Several simple facts explain why the tsunami led to a peace agreement on Sumatra but to 5 years of intense military conflict on Sri Lanka.

The Free Aceh Movement aimed to preserve several centuries of independence, while the Tamil Tigers sought independence after centuries of political and administrative unity.

Sumatra is almost entirely a religiously homogeneous Muslim region, whereas on Sri Lanka, 70% of the population were Sinhalese Buddhists, which makes the Tamil population a religious minority.

The Indonesian government was inclined to pursue peace to regain stability. In contrast, the Sinhalese nationalist government in Sri Lanka capitalized on the aftermath of the tsunami in its fight against the Tamil Tigers, as the tsunami mainly affected Tamil-controlled areas.

Indeed, another factor to consider is that Indonesia is incomparably larger than Sri Lanka (which is a territory comparable to that of Latvia or Lithuania). Therefore, Indonesia could afford to grant some form of autonomy to the province of Aceh in the interest of peace. Sri Lanka, on the other hand, was considered too small an island to be divided, and thus the official government sought to maintain its integrity.

Additionally, the distribution of humanitarian aid was much simpler in Sumatra than in Sri Lanka. The conflict in Aceh focused on the eastern coast of the province, while the tsunami mainly affected the western and northern coasts. Thus, humanitarian aid and military conflict did not necessarily intersect. In Sri Lanka, the majority of the devastated area was under the control of the Tamil Tigers.

Did So Many People Have to Die?

As we have already mentioned, it is estimated that almost 230,000 people died due to the tsunami. Certainly, the number of victims could have been significantly lower. An example is the communities on small islands, where stories of tsunamis have been passed down for many generations, and the islanders knew what to do in case of signs of an approaching tsunami. Thanks to this, thousands of people were saved.

The situation was different on larger islands and overall in the countries in the affected region. After the tsunami in 1883, neither politicians nor local authorities addressed the possibility of another tsunami. As a result, there was a lack of basic general education and awareness among the population about the possibility of a tragic natural phenomenon, and there was no tsunami warning system in the Indian Ocean. Despite all the tragic experiences, especially in the Pacific Ocean, and the existence of the Pacific Tsunami Warning Center, despite frequent submarine earthquakes in the subduction zone. Ignorance and underestimation of education had tragic consequences on a scale that the authorities couldn't even imagine.

If there had been a tsunami warning system in the Indian Ocean in 2004, along with better communication networks and basic education for the public about potential natural phenomena, leading to public preparedness for appropriate response and behaviour, it is very likely that at least tens of thousands of lives could have been saved.

It wasn't until the tragedy in December 2004 that countries in the Indian Ocean region established national tsunami warning centres. There was also an improvement in educating the public about the danger of tsunamis.

Tōhoku Earthquake and Tsunami in 2011

The ability or willingness to learn from ancient and recent history was sadly lacking in no small measure before the great earthquake and catastrophic tsunami that occurred on 11 March 2011 at 12:46 local time east of the Tōhoku region in the northern part of the island of Honshu. (The geographical area of Tōhoku includes, from north to south, the prefectures of Aomori, Akita, Iwate, Yamagata, Miyagi and Fukushima.)

The Mw 9.0 earthquake originated in a subduction zone where the Pacific Plate dips beneath the Okhotsk microplate which is a kind of Western peninsula of the North American Plate in an approximately western direction. At the latitude of the Tōhoku region, it is really fast—more than 8 cm/year. The line of a plate contact at the bottom of the Pacific Ocean is the Japan Trench, which reaches depths of up to 8 km. From the north, where it borders the Kuril-Kamchatka Trench, it extends approximately 800 km south to the Izu-Ogasawara Trench. The hypocentre of the earthquake was at a depth of about 20–30 km. The size of the ruptured part of the fault has been estimated in various analyses to be in the range of approximately 400–600 km in the horizontal direction and 150–220 km in the dip direction of the fault. The largest slip is estimated to have reached a value of about 50 m. The earthquake is estimated to have released 10^{17}–10^{18} J of energy. The maximum recorded acceleration of the seismic motion was up to 26.999 m/s^2 (i.e., 2.75 g) in Miyagi Prefecture.

After the earthquake, several thousand aftershocks occurred. The largest ones were Mw 7.9 on 11 March 2011 at 1:15 PM, Mw 7.1 on 11 March 2011 at 1:25 PM, and Mw 7.1 on 7 April 2011 at 9:32 PM.

The earthquake itself killed 1475 people, injured 6157 people, destroyed 2473 houses and damaged 5168 houses. Damage was estimated at 4.4 billion US dollars.

Most notably, however, the earthquake caused a huge tsunami, the scale and devastation of which we could watch almost online via television and the internet. Together, the earthquake and tsunami caused the deaths of 18,428 people, injured 6167 people, destroyed 123,661 houses and damaged 280,920 houses. The damage has been estimated at over 220 billion US dollars.

The Tōhoku tsunami caused a disaster at the Fukushima Daiichi nuclear power plant in Ōkuma, Fukushima Prefecture. We will write more about the disaster below.

Was It Possible to Expect an Earthquake with Mw 9.1 and a Devastating Tsunami in This Region?

Let's recall the previous earthquakes in the region from north to south: Sanriku M 8.5 in 1896, M 7.4 in 1901, M 7.6 in 1931, M 7.6 in 1933; Miyagi M 7.4 in 1897, M 7.4 in 1936, M 7.4 in 1978, M 7.2 in 2005; Fukushima Mw 7.4 in 1938, Mw 7.7 in 1938, and Mw 7.8 in 1938. For comparison, let's note that the Sanriku earthquake with a magnitude of 8.5 in 1896 released (very roughly) about 8 times less energy than the earthquake on 11 March.

It seems that not only many Japanese seismologists were greatly surprised by the magnitude of the earthquake on 11 March. Several were expecting "only" an earthquake in the magnitude range of 7 to 8. Professor Hiroo Kanamori stated in 2011 that the recent earthquakes had released so much accumulated stress that it was difficult to imagine where such a large earthquake could have been prepared. As we will mention in the chapter on earthquakes in Turkey, the earthquakes in February 2023 led to a similar feeling among Turkish seismologists.

Was this feeling or even assumption justified?

In the probabilistic seismic hazard analysis, it is never assumed that the magnitude of the largest earthquake over the entire documented period in the past (at least hundreds of years) is the maximum magnitude that can be expected. This is because the largest earthquakes can accumulate stress over longer periods, even hundreds or thousands of years. In hazard analysis, one of the possibilities considered, with appropriately chosen weight, is a magnitude up to one unit larger than the magnitude of the largest documented earthquake in the respective area. This means an earthquake where approximately 30 times more energy could be released. Of course, everything has its physical limits, and so far, as we have already stated, the values of material parameters inside the Earth, Earth rheology, and spatial heterogeneity inside the Earth probably do not allow for an earthquake larger than Mw 9.5, which occurred in Chile in 1960.

However, excluding the possibility of such a large earthquake in a subduction zone near Japan, with a plate subduction rate exceeding 8 cm/year, appears problematic after the Mw 9.1 earthquake off the coast of Sumatra in 2004. The Earth's interior is too materially and geometrically complex for us to rely on regularities in time, space, and magnitude. Especially when it comes to data for determining the level of seismic hazard or the possibility of devastating tsunamis in subduction zones.

There are also other important circumstances to consider.

It has been well known for quite some time that foreshocks occur in this subduction zone in the case of 50–70% of major earthquakes approximately 2–3 days before the main shock. On 9 and 10 March, hundreds of small earthquakes occurred in the area. The increase in the number of earthquakes was sudden and noticeable. The earthquake on 9 March 2011, at 9:45 local time, had a magnitude of Mw 7.2. As later revealed, it was the largest foreshock, and its hypocentre was less than 50 km away from the hypocentre of the main earthquake with a magnitude of Mw 9.1. On the day of the main earthquake, there were also three foreshocks with Mw > 6. The question arises whether it was reasonable and justified to issue at least a warning about the possibility of a major earthquake on 10 March? If the warning was not issued, was it because the relevant seismologists relied on the assumption that the earthquake with Mw 7.2 was the main earthquake? However, was relying on such an "optimistic" assumption reasonable? If the warning had been issued and the responsible authorities had responded adequately, several things could have been different.

Fortunately, the Japan Meteorological Agency issued at least a tsunami warning immediately after the main earthquake. This warning likely saved thousands of lives.

Professor Koji Minoura of Tōhoku University and his colleagues Fumihiko Imamura, Daisuke Sugawara, Yukio Kono, and Tomohiro Iwashita published an article titled *The 869 Jōgan tsunami deposit and recurrence interval of large-scale tsunami on the Pacific coast of northeast Japan* in the *Journal of Natural Disaster Science* in 2001. Professor Minoura and his colleagues analysed sediments in the plains of Sendai and Sōma and estimated the size of the earthquake that could have caused the tsunami. They concluded that in 869, an earthquake with a magnitude of approximately 8.3 occurred, rupturing a fault segment of 200×85 km^2. The earthquake triggered a tsunami that reached a height of 8 m on the coast. The tsunami penetrated up to 4 km inland. Professor Minoura and his colleagues estimated that the recurrence interval of a large tsunami in the Sendai area is between 800 to 1100 years. Given this interval and the fact that more than 1100 years have passed since the Jōgan tsunami, they emphasized that it is highly likely that the Sendai plain will be struck by a large tsunami. Using numerical simulations, they estimated that a tsunami similar to the Jōgan tsunami in 869 would penetrate approximately 2.5 to 3 km inland in the current coastal plain.

Several years before the earthquake on 11 March, seismologists began to agree that an event similar to the Jōgan earthquake could occur again. However, this scientific consensus, unfortunately, did not influence the relevant authorities and did not lead to a reassessment of seismic hazard and risk, improvement in tsunami preparedness, or a review of the resilience of the Fukushima Daiichi nuclear power plant.

Residents of Japan's northeast coast, largely low-lying flat areas, have long been well aware of their exposure to possible tsunamis. Also, as a result of earthquakes far from Japan. Recall that the 1960 earthquake in Chile triggered a tsunami that spread throughout the Pacific Ocean and killed 139 people and caused considerable damage in Sanriku, the northernmost area of the island of Honshu, almost 17,000 km away. Following the earthquake, 10 m-high walls began to be built in the north-east of Japan to protect harbours against tsunamis.

However, the protection of the Fukushima Daiichi nuclear power plant did not take into account the possibility of a tsunami higher than 5.7 m. It is therefore not surprising that the 14 m high tsunami caused by the earthquake of 11 March 2011 had no problem knocking out the plant's backup diesel generators. However, this was not the only safety issue in the plant's operation.

We should mention that the Kashiwazaki-Kariwa nuclear power plant in Niigata Prefecture already had problems due to the Mw 6.6 earthquake on 16 July 2007. The design protection against the effects of the earthquake underestimated the level of seismic hazard by a factor of 2 to 3. Fortunately, the design safety margins were large enough to accommodate the excess load, leading only to minor damage and weak radioactive waste.

Fortunate Circumstance

The tsunami on 11 March 2011 penetrated up to 5 km inland and devastated the Tōhoku region (Fig. 9.6). Field surveys demonstrated that in Iwate Prefecture, the tsunami reached a run-up height of up to 38.9 m. The run-up height is the difference between the height of the location reached by the tsunami (inundation line) and the reference sea level. The tsunami caused damage amounting to 55 million US dollars even in distant California and six million US dollars in the fishing industry in Tongoy, Chile, over 16,500 km away. However, its most interesting characteristic is that it was actually significantly smaller than it could or should have been for an earthquake of such magnitude! So, even though it may seem strange at first glance, despite everything, the Japanese were fortunate in their misfortune.

Why? The size and area of the uplift of the ocean floor, which are crucial for the generation of a tsunami, depend on how the fault rupture propagates. In the case of the earthquake on 11 March 2011, a larger tsunami likely did not occur because the relatively moderate propagation of the rupture in shallower depths was preceded by a more energetically significant propagation of the

Fig. 9.6 Even this bizarre position of the ship was a consequence of the devastating tsunami during the Tōhoku earthquake in 2011. NOAA—public domain

rupture in deeper depths. These two modes of rupture propagation are likely a result of friction on the fault plane changing with depth.

Fukushima Nuclear Power Plant Accident

The earthquake itself caused damage, but together with the tsunami and the accident at the Fukushima Daiichi nuclear power plant, it was unprecedentedly devastating for a large area in the vicinity of the plant.

At the plant, 3 nuclear reactors that were in operation at the time were shut down immediately after the tremors were detected by seismometers. As the earthquake had damaged the equipment supplying the plant from the electricity grid, the plant remained de-energised and back-up diesel generators were started up to supply the electricity needed for the emergency operation of the reactors, their cooling in particular.

The tsunami reached the plant about 40 min after the earthquake occurred. The first wave crest was 4–5 m high and did not break the 5.7 m high tsunami barrier. About 10 min after the first wave crest, the second and largest wave crest arrived with a surge height of 14–15 m. This one easily overcame the barrier.

The water penetrated the underground spaces where most of the backup generators were stored and damaged them. The water also damaged several pieces of equipment needed to cool the reactors and the backup batteries that were supposed to provide emergency operation for 8 h. This left the three reactors, which were shut down during the earthquake, completely without electricity. Operators were not only unable to control the reactors, but also had no information on their status. External mobile power generators could not reach the plant fast enough due to the damage to the roads caused by the earthquake and tsunami. When they finally did reach the plant, it was not possible to connect them quickly due to the damage caused by the tsunami. The nuclear fuel in the first reactor melted down the very next day, 12 March. That evening, the inhabitants of areas within a 20 km radius of the plant were evacuated. Over the following days, fuel meltdowns occurred in two other nuclear reactors. A month later, some villages beyond the 30 km radius were also evacuated.

However, it was not until June that the Japanese government released information that all three reactors had in fact melted down and that they were even considering evacuating Tokyo. The gradual informing was perhaps intended to prevent panic.

By May 2012, all 54 reactors in Japan had been decommissioned and were due to undergo a full inspection—making Japan nuclear-free for the first time since the 1970s.

The Fukushima Daiichi nuclear power plant has had its own history of shortcomings and neglect of risks. In previous years, they had already neglected and obfuscated the existence of cracks in the reactor cores, and had been cutting costs by reducing reactor inspections. The main problem was that the relocation of the diesel generators to a higher floor was not implemented, and a higher tsunami barrier was not built (as has been done at newer nuclear power plants in the region), even though the International Atomic Energy Agency had warned of these deficiencies.

After the Sumatra-Andaman earthquake, seismologists were the first to take notice and began to wonder whether a similar disaster could indeed befall Japan directly. In 2008, the power plant operator (TEPCO, Tokyo Electric Power Company) was forced by new research to consider the threat of an earthquake comparable to the Sanriku earthquake of 1896. According to calculations, such an earthquake could generate a more than 10–15 m high tsunami. This seemed too much for the plant operator, so he decided to ignore the warning as unrealistic. Nor did they respond to the warning of a potential earthquake similar to the one in 869. It was a "too historic" earthquake and did not fit the current conditions. TEPCO was unwilling to invest what was undoubtedly a very large sum of money in improving anti-seismic protection because of a potential earthquake that no one could quantify in advance.

After the accident, an unprecedented wave of public opinion against nuclear power surged. Public polls showed that two-thirds of respondents were in favour of Japan abandoning nuclear power altogether and that the population distrusted the government's safety measures. But there was nothing the government could do—Japan is a highly industrialised country and suddenly abandoning nuclear power would be a major problem. The plants had to be brought back on line, but under the condition of a new energy plan.

8% of Japan was contaminated with varying degrees of radiation, affecting approximately 160,000 Japanese who were evacuated to new areas. In 2013, it was reported that 1539 people had died, almost as many as a result of the earthquake, mainly due to the stress caused by the nuclear accident on residents in Fukushima Prefecture. Many of them suffered ill-health as a result of the evacuation conditions they lived in for months after the accident, while others were unable to cope psychologically and committed suicide.

Tens of millions of tons of contaminated land, including farmland, have been slated for disposal. Signs of radioactive waste intrusion have been found in drinking water. Contaminated food has been found in many places in

Japan. However, due to the extensive checks of contamination of food and drinking water, the radiological impact of the nuclear accident on the population has been minimised.

In 2012, Japan's first national "green" party, Midori no Tō, was founded, determined to abandon nuclear power plants as an energy source altogether.

In response to the Fukushima nuclear accident, Germany announced that it would completely abandon nuclear power by 2022. As of 16 April 2023, Germany does not produce electricity using nuclear power plants.

What's Next

One possible effect of the earthquake is the relaxation of stress off the coast of northeastern Honshu may have increased the stress in the southern part of the subduction zone, and thus increased the probability of a large earthquake closer to Tokyo.

The 2004 and 2011 earthquakes and tsunamis strongly indicate that the level of seismic and tsunami hazard in all subduction zones needs to be reassessed. It is also true for them that large earthquakes capable of producing devastating tsunamis can take a relatively long time to prepare. For example, longer than is well documented historically.

10

Turkey: Tragic Surprises and Tense Anticipation

Predicting Earthquakes

We could save thousands of human lives if we could predict the time and location of an earthquake that would cause strong seismic motions in densely populated areas of the Earth's surface. For example, if we could do this 24 h in advance, it would be possible to evacuate residents, halt critical operations, and shut down engineering networks. If we could do this 1 h in advance, people could at least exit buildings and gather in places that would not be threatened by collapsing buildings and falling objects. There would still be time to stop or limit some critical operations and shut down engineering networks. We are considering the real world with both developed and less developed countries. And therefore, the real (non)resistance and (poor)quality of constructions.

The ideal solution would be if we could predict earthquakes (time, location, magnitude) and only construct residential buildings, dams, power plants, factories, transportation routes, and engineering networks that could withstand the scientifically estimated maximum seismic motions. Of course, even in such a scenario, it would be reasonable to evacuate residents and implement preventive technical measures in the event of the largest earthquakes. After the earthquake, residents should be able to return to their homes, and power plants and factories could resume operations.

If we cannot predict when and where an earthquake will occur, seismologists should sufficiently accurately estimate the maximum values of important characteristics of seismic motions at a given location during future earthquakes. Designers should design structures accordingly, and builders should construct them diligently according to the construction project.

A Very Brief Period of Optimism

A period of optimism reigned especially in China after the magnitude 7 earthquake that struck the city of Haicheng (about 550 km from the centre of Beijing) at 19:36 local time on 4 February 1975. More than a month before the earthquake, changes in the height of the Earth's surface and the level of the water table had been observed. The number of cases of anomalous animal behaviour gradually increased. The number of weak earthquakes in the area also began to increase. Eventually, the sudden increase in the number and magnitude of earthquakes led the authorities to order evacuations. It is estimated that the evacuation saved at least 100,000 lives. Meanwhile, it is estimated that up to 90% of the buildings in the city were severely damaged or destroyed by the earthquake.

China has initiated a fairly extensive programme of monitoring signs of an impending earthquake, including anomalous animal behaviour.

Sobering Up from Optimism

The sobering up from optimism came sooner than anyone would have expected. On 28 July 1976, at 3:42 local time, an earthquake of magnitude Mw 7.8 occurred with a hypocentre about 15 km below the southern part of the city of Tangshan (the centre of the city is about 150 km from the centre of Beijing). No precursors, which led to the evacuation of Haicheng a year and a half ago. Official Chinese sources said that about 240,000 people lost their lives and about 160,000 were injured. Independent estimates put the death toll as high as 655,000.

Numerous earthquakes followed in different parts of the world, surprising either in size or in their effects. The most infamous are the 17 January 1994 earthquake of Mw 6.7, which originated on the previously unknown Northridge fault in California, and the large 17 January 1995 earthquake of Mw 6.9 in the Hanshin area of Japan, which originated on a fault to which seismologists had paid no attention. Both earthquakes caused unprecedented damage in their respective countries—seemingly at odds with the magnitudes of the earthquakes. However, as we know, the effects of earthquakes are not simply correlated with earthquake magnitude. Local conditions and the type and quality of structures also contribute significantly.

Both earthquakes have ominously highlighted a crucial aspect of densely populated areas in seismically active regions of the world: even relatively moderate earthquakes can cause significant tragedy and damage. This was perhaps most tragically confirmed by the Mw 7.0 earthquake that struck Haiti on 12 January 2010. The earthquake caused the death of at least 160,000 people and left approximately one million people homeless. It was not the earthquake that killed, but the incredibly shaky dwellings.

The effects of earthquakes were greater than predicted by seismic hazard analyses also for the 12 May 2008 earthquakes in the Sichuan region of China and the 11 March 2011 earthquakes in the Tōhoku region of Japan.

The tragedy of 6 February 2023 in Turkey and Syria is the most recent example of what can occur when seismic hazard is not properly accounted for by the building design and construction quality.

Finally, let us recall the alarming finding of the Swiss-American seismologist Max Wyss, who looked at the death toll from earthquakes between the years 2000 and 2010, including only those earthquakes that each killed more than 1000 people. There were 12 such earthquakes. Max Wyss was surprised to find that the actual number of victims of these earthquakes was 160 times higher than the number based on the probabilistic estimates.

Tectonic Situation in Turkey

The complex spatiotemporal occurrence of earthquakes in the territory of Turkey is a consequence of a very intricate tectonic situation, which primarily involves the interplay of three major lithospheric plates the Arabian, African, and Eurasian plates and the smaller Anatolian plate. The tectonic situation is illustrated in a very simplified way in the figure (Fig. 10.1).

The African lithospheric plate moves approximately northward at a speed of about 1 cm/year relative to the Anatolian plate. In its western part, it subducts beneath it, while in the eastern part, their contact is not well understood. The Arabian plate predominantly moves northward at a speed of approximately 2.4 cm/year relative to the Eurasian plate, although some experts also suggest a slight counterclockwise rotation. The Eurasian plate moves predominantly in an east-southeast direction relative to the Anatolian plate. The convergence of the African and Arabian plates towards the Eurasian plate causes the Anatolian plate to move westward or southwestward. This motion also occurs through motions along the North Anatolian Fault and the

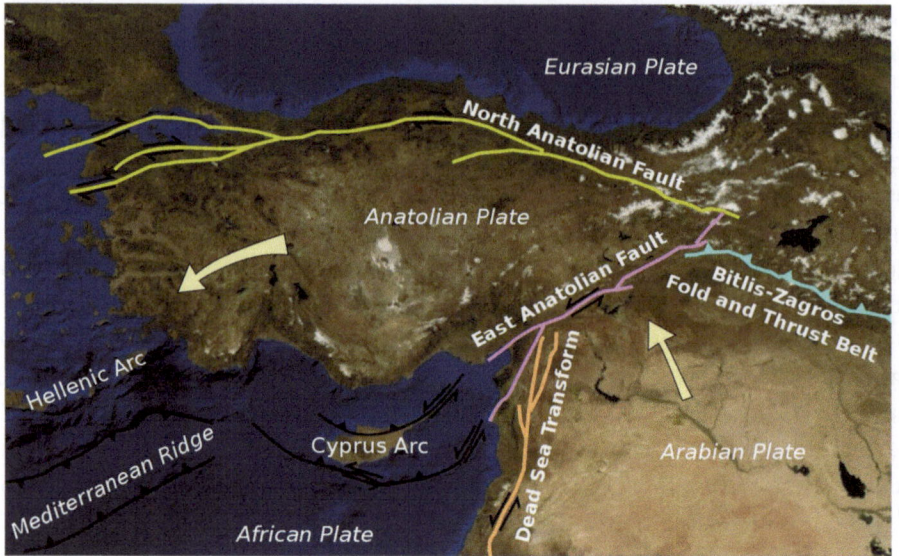

Fig. 10.1 The tectonic situation of Turkey and the occurrence of earthquakes on the territory of Turkey are mainly due to the mutual movement of the three large lithospheric plates—the Arabian, African and Eurasian—and the smaller Anatolian plate. The tragic earthquakes of February 2023 originated in the East Anatolian fault zone. Wikipedia.org—public domain

East Anatolian Fault. It is evident that the North Anatolian Fault is predominantly a right-lateral fault (similar to the San Andreas Fault in California), while the East Anatolian Fault is predominantly left-lateral.

The North Anatolian Fault is approximately 1200–1500 km long and shares many similarities with the San Andreas Fault in California. It extends from the east, from the triple junction of Karliova in northeastern Turkey all the way to the Aegean Sea. In the west it crosses the Marmara Sea and passes as close as about 20 km from Istanbul. The Anatolian plate moves along the North Anatolian Fault to the west relative to the Eurasian plate at a speed of approximately 2.5 cm/year.

The East Anatolian Fault, between the Anatolian and Arabian plates, is more than 500 km long and stretches from the triple junction of Karliova to the northeastern tip of the African plate, where it meets the Dead Sea Fault. In the northeast, the Anatolian plate moves relative to the Arabian plate at a speed of approximately 10 mm/year, while in the southwest, it moves at a speed of only about 1–4 mm/year.

The Dead Sea Fault is the boundary between the African and Arabian plates, and due to the faster motion of the Arabian plate to the north, it is a left-lateral fault.

Although this all may not sound simple, the reality is even more complex. The North Anatolian and East Anatolian faults are dominant ones, but there are dozens of active faults in the territory of Turkey. These include secondary faults in the immediate vicinity of the main faults, as well as faults within the Anatolian plate itself.

Antakya (Antioch), Years 115 and 526

The ancient Roman historian Cassius Dio (163–165—after 229 AD) addresses the earthquake of 13 December 115 in his Roman History. Antioch was a city with approximately 100,000 to 300,000 inhabitants at that time. During the winter of 115, the Roman emperor Trajan (reigned 98–117 AD) stayed in the city with his army during his expedition to the Near East. Additionally, due to various legal, political, and commercial reasons, a large number of foreigners also resided in the city. It is not clear by how much this number of people residing in the city at that time increased, but several authors suggest that throughout the region up to 260,000 people may have perished in the earthquake. A large number of people were seriously injured. Emperor Trajan spent the rest of his stay in a tent in the open space of the hippodrome.

According to historical documents, the earthquake in May of 526 was also tragic. The earthquake is mentioned in the chronicle of the Byzantine chronicler John Malalas (c. 491–578 AD). As a result of the earthquake and the ensuing fire that engulfed the city, reportedly 250,000 people lost their lives. Emperor Justin I (reigned 518–527 AD) and his nephew Emperor Justinian I (reigned 527–565 AD) made great efforts to rebuild the city, focusing on reconstructing Christian buildings. However, the enormous loss of life and extensive material damage caused by the earthquake significantly contributed to the long-term decline of Antioch, which in previous centuries had been compared to cities like Alexandria, Athens, Rome, and Constantinople.

Millenia of Earthquakes in Istanbul

Thanks to the long history of human settlement on the Anatolian Peninsula, particularly in the area of present-day Istanbul, we can study almost two millennia of repeated earthquakes and their impact on civilization.

In the liturgical calendar, after the founding of Constantinople (present-day Istanbul) in 330, every major earthquake that struck the city was documented. On anniversaries, the city's inhabitants would then commemorate the tragedies with a public procession and Mass. The material and social impact of earthquakes and other frequent disasters such as plagues, famines, or fires had a direct influence on the structure of the city's settlement and its surroundings.

The first documented earthquake struck Constantinople in 358, less than 30 years after its founding. The earthquake also damaged the city of Nicomedia (today's Izmit).

During an earthquake during the reign of Emperor Theodosius II (reigned 402–450 AD), Nicomedia was again damaged, and 57 towers on the Theodosian Walls in Constantinople collapsed. This significantly weakened the fortress system of Constantinople at a time when the dreaded Attila the Hun's armies were approaching the city walls. Under this threat, the city's inhabitants were mobilized and the walls were reconstructed within 3 months.

Constantinople was hit by major earthquakes at least once a century (and that was during the quieter centuries). The city was repeatedly built and reconstructed. Witness of the centuries of earthquakes is Hagia Sophia. The church was built in the fourth century. However, its present form is markedly different from the original one. In fact, the building has been reconstructed several times, also due to earthquakes, and each reconstruction has brought with it a new architectural element. In the sixth century, the builders and engineers responsible for the reconstruction of Hagia Sophia added flint or volcanic ash to the mortar. These ingredients reacted with water and the result was a building material that was more flexible and could absorb some of the seismic energy (the energy of mechanical oscillations).

Earthquake in 557 The earthquake was part of a series of both major and minor earthquakes that struck Constantinople in the middle of the sixth century. The earthquake hit Constantinople on 14 December. Many homes, churches, baths collapsed, and Hagia Sophia was damaged, likely leading to its collapse a year later due to ongoing aftershocks. It was rebuilt on the orders of the ruler Justinian I, using different materials. Ultimately, Hagia Sophia acquired a new form and its present height. In 559, Slavic and Hunnic armies advanced as far as the earthquake-damaged Anastasian Walls, about 65 km from Constantinople.

A dramatic testimony of the earthquake is provided by Agathias Scholasticus (c. 530–between 582–594 AD), who served as a historian at the court of Emperor Justinian I between 552–558. He describes the earthquake in the fifth book of his work *On the Reign of Justinian*. In this work, he writes that Constantinople was almost levelled to the ground by the earthquake. He describes the tremors as spasms of incomparable magnitude and duration, the horror of which was further intensified by what followed after the earthquake itself.

The earthquake struck sometime before midnight when most people were asleep in their homes. The tremors woke people up, and panic erupted; the city resounded with cries, prayers, and dreadful loud sounds emanating from Earth's interior. People rushed into the streets to save themselves, but Constantinople was so densely covered with buildings that it was not easy to find safety in open spaces.

Agathias goes on to describe that the air was full of dust and smoke, and freezing rain and snow began to fall on the crowds of people gathered in the streets. However, people were afraid to go back into the buildings. Only some took refuge from the cold in the churches where they prayed. In addition to the description of the collapsing buildings, which is often difficult to believe in its dramatic nature, Agathias provides an interesting probe into the people's actions during that tragic night. He writes that a large number of women of all social classes went beyond social decorum and consorted even with men of lower status. Slaves rebelled in fear against their masters and sought refuge in places of worship. Under the threat of the natural disaster, the boundaries and hierarchy that society had established were, for a while, completely wiped away.

When the sun rose in the morning, people joined their loved ones in joy, hugging, kissing and crying with relief.

The tremors lasted for several days after the earthquake. Although they were shorter and milder, they still prevented the city from returning to everyday life as quickly as possible. Agathias speaks of fantastic tales of the end of the world that circulated through the crowds, supported by various charlatans and self-proclaimed prophets who were fooling the demoralised population.

Emperor Justinian I showed respect to the victims by not wearing the imperial crown for 40 days. The earthquake was commemorated with an annual liturgy in Constantinople. At the same time, the emperor warned the people that earthquakes were a manifestation of divine wrath against loose social morals.

Earthquake in 1509   In 1453 Constantinople was conquered by the Ottoman Turks. Sultan Mehmed II (reigned 1444–1446 and 1451–1481) decided to renovate and repopulate the city. During the 50 years after the conquest, the new capital of the Ottoman Empire was free from earthquakes.

However, on 10 September 1509, during the reign of Bayezid II (reigned 1481–1512), an earthquake so terrible struck that it earned the nickname 'Minor Judgement Day' (Fig. 10.2), referring to the Islamic tradition where a tragic earthquake heralds the Last Judgement. The earthquake and the tsunami it triggered were seen at the time as punishment for political, military and moral failures.

4000–5000, according to some sources even up to 13,000 people lost their lives and 10,000 were injured. The Sultan's Topkapi Palace was also affected, and the Sultan himself was only saved from death by being outside his bedroom, which collapsed as a result of the earthquake. For a few days afterwards he lived in a tent in the palace gardens, then temporarily moved to Edirne. According to the Sultan, the earthquake was divine punishment—but punishment for the offenses and mistakes of his political dignitaries.

According to historical sources, not a single building remained intact. The remains of the walls of Constantine the Great collapsed. More than a hundred

Fig. 10.2 The earthquake of 10 September 1509 hit Constantinople so devastatingly that it was nicknamed 'Minor Judgement Day'. Based on an engraving by an unknown artist. © Ladislav Csurma, 2023. All rights reserved

mosques were damaged. The minaret that was added to Hagia Sophia after its conversion from a Christian church to a mosque following the Turkish conquest of Constantinople was also razed to the ground and the plaster that covered the Christian mosaics fell away.

A damage assessment was ordered after the earthquake. 66,000 workers, carpenters and stonemasons, from various parts of the empire were employed for the reconstruction, and taxes were raised the following year to cover the reconstruction.

More than a thousand dwellings reportedly fell, leaving many people homeless. At the same time, because of the 45-day-long aftershocks, they were unable to start reconstruction. Warehouses and food stores, mills, shops and aqueducts were also damaged, resulting in food shortages and limited access to drinking water.

Earthquake in 1766 In the summer of 1766, Istanbul was struck by further tragic earthquakes. The earthquake on 22 May (Fig. 10.3) triggered a tsunami that affected the Bosporus and the Gulf of Gemlik (south-east of Marmara Sea). According to preserved documents, the earthquake claimed 5000 lives

Fig. 10.3 The earthquake on 22 May 1766, was also tragic for Istanbul. Based on the original engraving by William Henry Bartlett from the nineteenth century. © Ladislav Csurma, 2023. All rights reserved

and caused such extensive panic that the Sultan ordered responsible authorities to patrol the streets. The 5 August 1766 Marmara earthquake even further exacerbated the situation in Istanbul caused by the earthquake in May.

In the case of the 22 May earthquake, certain parallels can be drawn with the Lisbon earthquake that occurred 11 years earlier. In Istanbul, the earthquake occurred at 5 AM, shortly after the morning prayers, which are performed at sunrise. Fortunately, the mosques were already empty at that time—unlike the churches in Lisbon, which were full of believers attending the early morning Mass on 1 November 1755, All Saints' Day.

A significant factor in comparing the consequences of the Lisbon and Istanbul earthquakes is the contemporary architecture. Both in Europe and in the Ottoman Empire, the half-timbering construction of buildings had had been widespread for centuries. However, these architectural elements were not originally intended as seismic-resistant measures. Such aspect of half-timbering construction gained significance when better resistance to tremors was repeatedly observed in buildings constructed in this manner in seismically active areas compared to other types of structures. According to European travellers, timber-framed houses were very common in Istanbul. While this construction was advantageous in the event of an earthquake, its downside lay in the frequent fires that engulfed Istanbul continuously for centuries.

From the perspective of urban development, Lisbon and Istanbul adopted very different measures after the earthquakes. The original Lisbon was largely rebuilt, creating an opportunity to introduce new building standards. The city's reconstruction also included the construction of the Baixa Pombalina district, known for its distinctive architecture. It features a rectilinear layout with strict regulations for the construction of half-timbered buildings. The wooden construction of buildings—gaiola—indeed resembles "cages" filled with building material. Architectural decisions were based on accounting for the flexibility of the construction material.

Istanbul opted for a different tactic. There was no scientific approach aimed at stabilizing newly constructed buildings. Most attention was given to the reconstruction of historically and religiously significant buildings. Residents of residential buildings affected by the earthquake repaired their homes hastily, with little attention to their strength or resistance.

Earthquake of 1894 When an earthquake with a magnitude of 7.3 struck on 10 July 1894, Istanbul was certainly a modern city but not earthquake resistant. It is estimated that approximately 10,000 buildings were damaged as a result of this earthquake and its aftershocks. The railway network was damaged, and only one telegraph line survived in the city. Water pipes cracked, leading to contamination of drinking water.

The earthquake had a huge impact on Istanbul society. Out of fear and due to constant aftershocks, residents began to create dwellings in public open spaces, squares, cemeteries, and public parks.

It was this earthquake that sparked scientific interest in seismic activity along the Anatolian faults. It was supported by a new scientific discipline—seismology—which had been finding its place in intellectual circles since the mid-nineteenth century. In response to the earthquake, Sultan Abdülhamid (reigned 1876–1909) asked Demetrios Eginitis, the director of the Greek National Observatory (1862–1934), to come to Istanbul and prepare a report on the consequences of the earthquake. Ottoman ambassadors were to inform the Sultan about the state of seismological research in European countries. Soon after, the first Ottoman state-funded observatory was established. Its director was the Italian seismologist Giovanni Agamennone (1858–1949). Agamennone formed a team and trained the first generation of Ottoman seismologists.

Aleppo 1822

On 13 August 1822, the northwestern part of present-day Syria and the southern province of Turkey was struck by an earthquake with an estimated magnitude of 7.0–7.4. The epicentre is estimated to have been around the Syrian town of Afrin, approximately 45 km from the historically significant city of Aleppo (nowadays Halab).

The earthquake occurred during the summer evening of 1822, at approximately 9 PM local time, when many people were on the roofs or courtyards of their houses. Many people were therefore in the open, and a number of them managed to save themselves. Greek earthquake engineer and seismologist Nicholas Ambraseys estimated the death toll at between 30,000 and 60,000. It seems well documented that 7000 deaths occurred within the city walls itself.

Every part of the city bore the marks of the earthquake, and life in the city came to a halt for a time. The tragedy sometimes lies not in the earthquake itself and the immediate damages it inflicts but in the unpredictable and sometimes exceedingly surprising long-term secondary consequences. As a result of the earthquake and its aftershocks, the frightened population was forced to abandon their dwellings in the city and retreat to wooden huts and tents in the outskirts. Many remained there for months without a chance to begin the reconstruction of their homes. The bold people who returned to the city soon after often perished during the aftershocks.

The governor and judge also left the city, and according to records, municipal courts did not function for 40 days. In the devastated city, as well as in provisional dwellings in the surrounding areas, crime rates increased, leading to patrols being ordered.

Although the stone architecture of the city had taken into account centuries of experience with earthquakes, the earthquake in 1822 was unprecedentedly large. As a result of the earthquake, the residents of Aleppo had to resort to various short-term and long-term measures that affected the development of urban architecture. One of these measures was the introduction of the half-timber construction, in which the interior of the walls was filled with material from the ruins of buildings.

The European powers, especially the French, English, and Dutch, maintained direct contact with Aleppo. They appointed consuls from the ranks of Aleppo's Jewish and Christian merchant families, who informed their embassies about the situation's development.

The day after the earthquake, foreign merchants and European consuls left the city and managed to obtain a long-term lease of land outside the city. They built wooden buildings, residences, and even a chapel there, and by the end of the nineteenth century, this conglomerate of buildings became a foreign quarter of Aleppo, known as the 'petite ville franque', mainly inhabited by European immigrants. In this multicultural environment, a part of the city took on a completely new form, characterized by the blending of local Middle Eastern traditions and various European influences. The emergence of this quarter was also one of the effects of the earthquake on the architectural development of the city.

After the earthquake, the already ongoing process of concentration of real estate owned by wealthy and influential families was accelerated. These families were often of Italian-Jewish origin and had been in Aleppo for a long time, sometimes for several generations. These wealthy merchant families invested in the reconstruction of the buildings, giving them a modern appearance typical of the nineteenth century. Thus, the earthquake was an opportunity for wealthy investors to buy up houses in Aleppo's commercial centre, accelerating the creation of a new commercial aristocracy in the city.

Earthquakes in Twentieth and Twenty-First Centuries

During this period, prior to the two earthquakes in February 2023, there were 18 earthquakes with magnitudes ranging from 7.0 to 7.9 in the territory of Turkey (Fig. 10.4). None of them occurred on the East Anatolian Fault. Eleven of them occurred on the North Anatolian Fault, with one on its southernmost branch—south of the Sea of Marmara. The most tragic was the earthquake on 26 December 1939, with a magnitude of 7.8 near Erzincan at the eastern end of the North Anatolian Fault. It killed over 32,000 people. Sixty years later, the earthquake on 17 August 1999, with a magnitude of 7.6 near Izmit, close to Istanbul, killed more than 17,000 people. Fortunately, in the case of the other earthquakes, the number of fatalities was lower, with fewer than 4000 deaths.

Erzincan and War Earthquakes

In the middle of a December night in 1939, the northeastern city of Erzincan was hit by a devastating earthquake. This earthquake was the beginning of a series of annual earthquakes of great intensity and consequence. As a result of the earthquakes of 1942, 1943 and 1944, the Turks experienced a somewhat different version of the World War II than the rest of the world. However, due

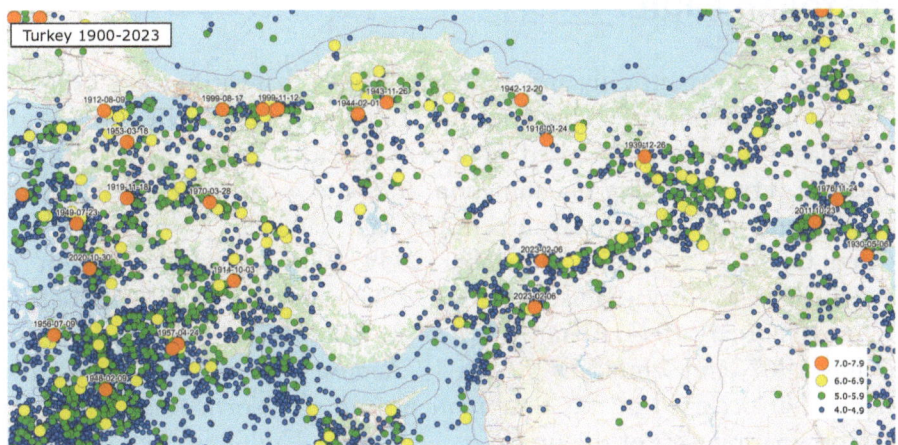

Fig. 10.4 The map of earthquake epicentres with a magnitude equal to or greater than 4 in the period 1900–2023 in the territory of Turkey and its surroundings. Wikipedia.org—public domain

to the political situation, Turkish seismologists were left to fend for themselves, and foreign experts were only able to reach Turkey after the end of the war.

However, it was precisely because of the unfortunate geopolitical situation that the US took an interest in the state of the Republic of Turkey, because there was a possibility that it would side with the enemy, as it did during the World War I. The communication between Turkey and the US on both scientific and diplomatic levels led to the establishment of technical cooperation which was important during the Cold War.

In response to the earthquake in Erzincan, the United States, particularly California, which had its own experience with devastating earthquakes, provided assistance to the Turkish government. This aid enabled Turkey to develop its own program for disaster prevention and mitigation, as well as to implement building standards. The seismic load in earthquakes is calculated using a principle very similar to that used in the USA. A map of seismic active zones with three levels of hazard was created.

In Turkey, earthquakes in the twentieth century, especially those in the 1940s, 1960s, and 1990s, claimed more than 90,000 lives, injured 175,000 people, and caused the collapse of 650,000 buildings. Building standards underwent reevaluation sporadically, approximately every decade, but especially after earthquakes that caused significant damage.

Migration of Large Earthquakes on the North Anatolian Fault

No fault is so geometrically and materially simple that earthquakes occur on it in a simple spatiotemporal arrangement or pattern. Nevertheless, the North Anatolian Fault is characterized by a very remarkable and, as we shall see shortly, a very disturbing phenomenon. It is also a consequence of the fact that the fault is nearly vertical and entirely right-lateral.

Erzincan M 7.8 on 26 December 1939, Niksar-Erbaa M 7.0 on 20 December 1942, Tosya-Ladik M 7.2 on 27 November 1943, Bolu-Gerede M 7.2 on 1 February 1944, Abant M 7.1 on 26 May 1957, Mudurnu M 7.1 on 22 July 1967, and Izmit Mw 7.6 on 17 August 1999 are the major earthquakes that occurred in this time sequence on the North Anatolian Fault, each originating west of the previous earthquake. Undoubtedly a significant migration of large earthquakes along a single fault (Fig. 10.5).

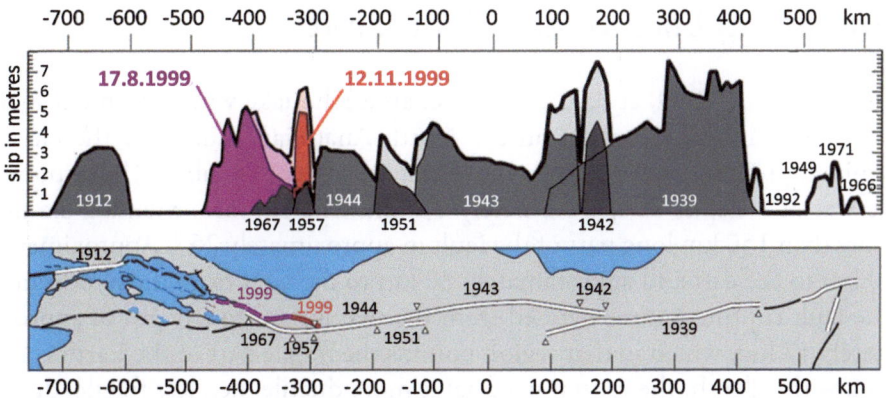

Fig. 10.5 The figure illustrates the migration of large earthquakes along the North Anatolian fault. It is very disturbing that the only segment of the fault that has not experienced a major earthquake in over a century is near Istanbul. Seismologists focusing on this area are concerned that the probability of an earthquake with a magnitude greater than 7 is quite high. An earthquake there could cause a disaster. Adapted with permission from Stein et al. (1997) [© Oxford University Press, 1997] and Armijo et al. (2000) [© The Authors, 2000]. All rights reserved

Also important is the fact that at a distance of about 150 km to the west from the western end of the ruptured segment of the fault in the Izmit Mw 7.6, 17 August 1999, earthquake, is the eastern end of the ruptured segment of the fault in the Mürefte M 7.3, 9 August 1912, earthquake. In other words, between the ruptured segments of the fault from the 1912 and 1999 earthquakes is a segment about 150 km long that has not experienced a major earthquake in more than a century. The zone extends from east to west and its shortest distance from Istanbul is only about 20 km in a southerly direction.

Obviously, there is a potentially huge problem here. Will the next big earthquake be in this seismically quiet zone south of Istanbul? If so, what can be expected? According to recent estimates by serious experts, an earthquake with a magnitude above 7.5 could cause the deaths of about 100,000 people. Seismologists and seismic engineers alike would love to be wrong in this case.

In thinking about the possible consequences of the feared earthquake, it helps to recall the aftermath of the 1999 earthquake near Izmit.

Unsettling Earthquake Near Izmit

On 17 August 1999, at 3:01 local time, an earthquake with a moment magnitude of Mw 7.6 occurred on the North Anatolian Fault near the city of Izmit, approximately 50 km from the eastern edge of Istanbul. The hypocentre was at a depth of approximately 15 km. The rupture extended along a more than 150 km long part of the fault in approximately 25 s. Approximately 70 km to the east and approximately 80 km to the west from the hypocentre. The fault ruptured area extended from the free surface to a depth of approximately 17 km, which in that region bounds the fragile part of the Earth's crust (below this depth, the crustal material is more ductile, i.e., less fragile, due to increased temperature). During the earthquake, the Anatolian Plate shifted approximately 3 m westwards on average relative to the Eurasian Plate on the ruptured part of the fault.

The earthquake killed more than 17,000 people, destroyed more than 110,000, and damaged more than 260,000 buildings. Entire apartment complexes collapsed in the most affected areas. More than 600,000 people were left homeless. The damage was estimated at 10–15 billion US dollars.

The reason for the high number of casualties was not the lack of knowledge or building standards. The reason was that, due to the huge migration of the population from the countryside to Istanbul in search of work, housing was built cheaply, and the main problem of developing countries in particular was fully manifested: corruption. Poorly paid building inspectors make a living by approving flimsy buildings that have a chance of collapsing even without earthquakes.

During the second half of the twentieth century, the population of Istanbul grew from approximately one million inhabitants to 8.5 million inhabitants. This rapid increase was made possible by the expansion of the labour market and urbanization. It meant an immediate need to expand the space. Therefore, illegal slums also started to appear on the outskirts of the city. Traditional brick buildings and buildings with wooden structure began to be increasingly replaced by high-rise concrete buildings. Building standards that take earthquake risk into account were first introduced in 1940 and have been updated every decade. The problem was, and ultimately still is, in the inconsistency in overseeing the application of these standards in practice. The constructions were and are often insufficiently strong, the materials are of poor quality, and the funds intended for construction are being stolen.

The view of the town of Gölcük, located not far from Izmit, under the easternmost tip of the Sea of Marmara, was bizarre. The mosque and minaret remained virtually undamaged amid the ruins of the apartment blocks.

In the town of Degirmendere near Gölcük, a massive landslide occurred as a result of the earthquake. An approximately 100 m long part of the coast sank into the sea. Hotels, shops and restaurants sank two stories below sea level.

A fire broke out at an oil refinery in Izmit, threatening to spread the flames further into the industrial part of the city, where water pipes had cracked as a result of the earthquake.

Foreign countries responded to the earthquake with immediate humanitarian and medical aid and efforts to prevent the uncontrolled oil leakage into the sea as quickly as possible.

Rescue operations after the earthquake were delayed due to damaged roads and damage to the highway connecting Istanbul and Ankara. Alternative routes were completely blocked by panicking people trying to connect with their loved ones. Looking for them in person was the only option since the phone connections were down.

After the earthquake, there was a justified wave of criticism of the quality of construction and the approach of the authorities. The strong criticism and the prediction of a 41% probability of an earthquake with a magnitude greater than 7.0 on a quiet segment of the North Anatolian fault south of Istanbul led the government and authorities to move forward with a vigorous effort to reinforce buildings in Istanbul and to crack down on corruption in the construction industry. A 20-year reconstruction programme was adopted, which was to require 400 billion US dollars. It is to be hoped that the objectives of the programme were successfully implemented.

Interesting Observations It is interesting that 2 days before the earthquake in Izmit in 1999, residents of the town of Degirmendere noticed a large number of dead crabs and jellyfish on the shore. The dead crabs and jellyfish could have died from radon released from the Earth's crust into the seawater. The unusual behaviour of fish, jumping above the water surface, was also observed.

During the earthquake, residents in the bay area observed fiery flashes and explosion sounds. This could be explained by the release of methane from the marshes in the bay due to friction during the rupture propagation on the fault.

Several days after the earthquake, flashes of strong light emanating from the sea were observed above the bay and the northeastern part of the Sea of Marmara. Similar flashes were also mentioned in connection with other earthquakes. They can even be seen in contemporary depictions of earth-quakes in the years 1509 and 1556.

Too Fast a Rupture?

In a simplified view, the North Anatolian fault is a vertical right-lateral fault. During the period in which the plates do not move along the fault itself due to friction (i.e., during the earthquake preparation period), tangential stress increases on the fault. The Eurasian plate tries to "pull" Anatolian plate to the right. This is the direction of the initial pre-earthquake stress on the fault. Of course, if I choose the Eurasian Plate as my observing site, the Anatolian Plate is exerting a traction on the Eurasian Plate trying to pull it to the right. So, the direction of the stress (we should correctly say the direction of the traction vector) depends on the chosen observing position. In any case, the direction of the tangential stress is horizontal, and the rupture could propagate from the hypocentre equally in both the westward and eastward directions if the conditions for propagation were the same in both directions. Often, however, in complex reality they are not and rupture propagation in one direction from the hypocentre may dominate.

In many large earthquakes on vertical right-lateral or left-lateral faults, the rupture predominantly propagates in one direction. One reason, for example, is that in the other direction there is not yet sufficient stress accumulated on the neighbouring fault segment after the previous earthquake.

In the 1970s, Robert Burridge and Lambert Ben Freund theoretically, and Joe Andrews, Keiti Aki, and Shamita Das numerically examined the propagation of ruptures on materially and stress-homogeneous transform faults (the San Andreas Fault and the North Anatolian Fault being good examples). They demonstrated that ruptures can propagate at a velocity slower than the velocity of Rayleigh waves (approximately 0.92 times the speed of S-waves, known as sub-Rayleigh velocity) or at the so-called "supershear" velocity, which is greater than the speed of S-waves but less than the speed of P-waves. In the case of "supershear," one can somewhat analogize it to supersonic propagation in the air. The velocity at which the rupture propagates depends on two factors. The first factor is the friction on the fault. The second factor is the magnitude of the initial stress on the fault, i.e., how prepared the fault is for an earthquake.

Velocities of rupture propagation were later determined for actual earthquakes recorded by a sufficient number and necessary configuration of seismic stations. In all cases, the rupture propagated at sub-Rayleigh velocities.

American seismologist Professor Ralph J. Archuleta, however, made a surprising finding when analysing an earthquake that originated on the Imperial fault in California just off the US—Mexico border on 15 October 1979 (Mw 6.5 Imperial Valley). The seismic station records could only be explained by a supershear rupture propagation.

In 2001, French seismologist Dr. Michel Bouchon and his colleagues published surprising result of their analysis of the earthquake in Izmit. The rupture propagated westward at sub-Rayleigh velocity, while eastward propagation occurred at a velocity greater than the speed of S-waves, specifically nearly 5 km/s. Similar propagation was also observed in the case of the Düzce earthquake of Mw 7.2 on 12 November 1999.

In the following years, supershear rupture propagation was identified in more than 15 earthquakes. For example, 2004 Mw 7.8 Denali, Alaska, or 2013 Mw 7.5 Craig, Alaska. In all cases, supershear rupture propagation occurred in the case of relatively simple fault geometry. The ruptures appear as unexpectedly thin lines on the free surface. It is worth noting that a rupture propagating at supershear velocity can cut the trunk of a tree.

During earthquakes with supershear rupture propagation, accelerations of seismic motion near faults are surprisingly low, seismic waves at higher frequencies are relatively weakly generated, and aftershocks are almost nonexistent. These characteristics are consistent with the relatively simple geometry of the fault, relatively homogeneous strength, and initial stress on the fault.

Precursors of an Upcoming Earthquake?

Laboratory and theoretical studies suggest that earthquakes are preceded by a phase of slip instability development during which slow slip occurs on the fault before rupture occurs. Michel Bouchon and his colleagues found that during the 44 minutes before the Izmit earthquake occurred, small seismic events originated at its hypocentre generating seismic waves recorded by seismic stations near the hypocentre. They identified 18 events. The smallest had a magnitude of 0.3, the largest 2.7. Very roughly, the largest released almost 4000 times more energy than the smallest. Michel Bouchon plotted the records of each event so that they all had the same largest amplitude. He was surprised to find that the records were virtually identical! Despite the range of magnitudes! Thus, a characteristic signal was repeatedly produced in the hypocentre before the large earthquake occurred.

The entire process can be roughly imagined, perhaps with squinted eyes, as the tensioning of a rope woven from steel wires at some depth beneath the

Earth's surface. The rope is subjected to pulling from both ends and slowly elongates due to the tension. At some place, for example, due to material heterogeneity, one wire breaks. Since the pulling tension does not cease (which corresponds to the fact that the plates are still moving), after some time, another wire or even several wires break. Each of these wire breaks represents an event that can be recorded on the Earth's surface by sensitive seismometers if they are close enough. Each of these wire breaks weakens the rope overall, so with continued tension, one can expect the next break to occur sooner. Eventually, the rope will completely break, and both of its parts will spring apart over a relatively large distance.

Does this mean that if such a development were to occur in the hypocentre of a future large earthquake, at least on the transform faults, and if there were sensitive seismic stations in a reasonable configuration near the hypocentre, it would be possible to detect an approaching earthquake with an advance of a few tens of minutes? This is an open question that would be good to answer as soon as possible.

Neither all the major faults hundreds of kilometres long near populated areas of the Earth's surface are monitored by seismic stations in such a way as to always and everywhere detect this final phase of preparation for a large earthquake. As we already know, there are secondary faults at major faults that may gradually become more frequent hosts of earthquakes than the original main fault. Even if we were to know all of them, monitoring all of them is currently impossible, at least for financial and logistical reasons. Thus, here we may recall the main task of seismologists, which is and will be to predict as accurately as possible what will happen at the site of interest during future earthquakes.

6 February 2023: Tragic Surprise on the East Anatolian Fault

Since 1999, Turkey has feared a major earthquake on a seismically quiet segment of the North Anatolian fault at a short distance south of Istanbul. On 6 February 2023, however, two earthquakes on the East Anatolian Fault surprised seismologists and the public at once.

At 4:17 AM local time, an earthquake with a moment magnitude of Mw 7.8 occurred about 37 km west-northwest of the city of Gaziantep, about 50 km from Turkey's border with Syria. This means the largest earthquake since the 1939 earthquake near Erzincan (which had the same magnitude).

The whole world was shocked from the first reports, which indicated that a catastrophe had occurred. Many know that aftershocks occur after such an earthquake. Indeed, aftershocks began practically immediately. However, nobody, including seismologists, anticipated that approximately 9 h later, at 13:24, another earthquake with a magnitude of 7.6 would strike approximately 95 km north-northeast of the epicentre of the earlier earthquake. Roughly, during the second earthquake, only about twice less energy was released compared to the first one. Therefore, one might speak of an earthquake doublet (Figs. 10.6, 10.7 and 10.8).

Within 24 h of the first earthquake, more than 570 aftershocks were recorded. The largest, with a magnitude of 6.7, occurred just 11 min after the earthquake. Typically, the maximum magnitude of aftershocks following large

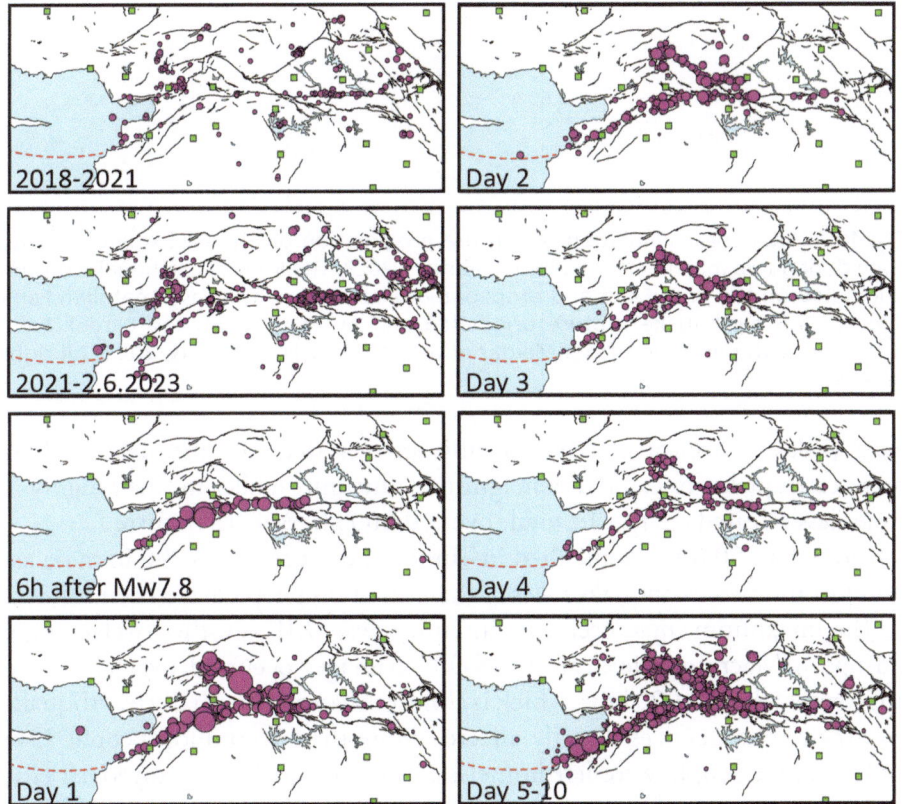

Fig. 10.6 Maps of epicentres (purple circles) of earthquakes in the East Anatolian fault zone at different time periods. Mw 7.8 indicates a tragic earthquake on 6 February 2023 at 4:17 local time. Green squares indicate cities. Based on Karabulut et al. (2023), © Geological Society of London, 2023. All rights reserved

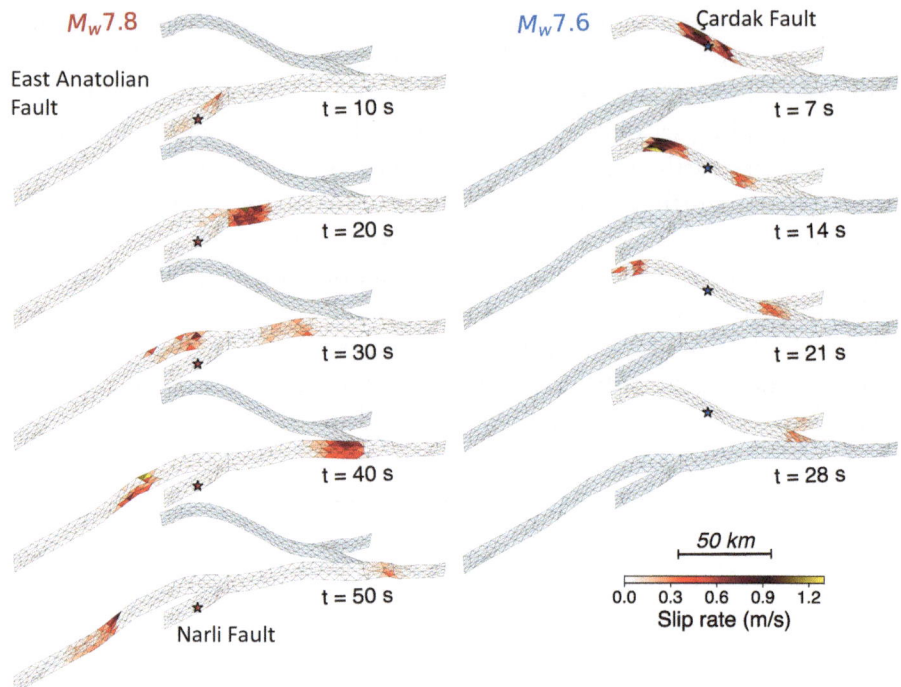

Fig. 10.7 Numerical simulation of rupture propagation on the East Anatolian Fault zone during two major earthquakes in February 2023. The Mw 7.8 earthquake rupture originated on the Narli Fault and propagated bilaterally on the East Anatolian Fault. The Mw 7.6 earthquake rupture originated and propagated on the Çardak Fault. Reprinted with permission from Melgar et al. (2023), © The Authors, 2023. All rights reserved

earthquakes is approximately one unit smaller than the magnitude of the main earthquake. Recall that a magnitude one unit smaller means roughly 32 times less released energy. In total, over 10,000 aftershocks occurred.

More than 50,000 people died in Turkey and more than 7200 in Syria as a result of the two earthquakes. More than 100,000 people have been injured.

The maximum macroseismic intensity was XII on the twelve-degree European Macroseismic Scale EMS-98. Damages occurred over an area of approximately 350,000 km², which is almost the area of Germany. Earthquakes and their consequences directly affected more than 16 million people. Over eight million people were left homeless. The earthquakes destroyed 12 cities and severely damaged several districts. More than 300,000 residential homes and family houses and over 150,000 commercial buildings were destroyed or severely damaged. Overall, the effects of the earthquakes were seen in

Fig. 10.8   A surprisingly thin surface trace of a Mw 7.8 earthquake rupture on a soft soil surface. Image taken by the Maxar satellite system. Maxar Open Data Program

approximately four million buildings. Direct damages were estimated at over 100 billion US dollars.

Seismology of an Earthquake Doublet

The Mw 7.8 earthquake began on the Narli Fault, an oblique fault to the East Anatolian Fault, at a depth of approximately 14 km. The epicentre was about 15 km east of the East Anatolian Fault. The rupture propagated approximately 50 km from the hypocentre towards the East Anatolian Fault. Preliminary analyses led some seismologists to consider the possibility that supershear propagation had already occurred on the Narli Fault. When the rupture reached the East Anatolian Fault, it triggered rupture propagation in both directions—southwest and northeast—initiating predominantly left-lateral horizontal slip. Eventually, it spread to a total length of 300 to 350 km, reaching a maximum velocity of approximately 3.2 km/s, which is approximately 90% of the S-wave speed in that area. Slip of 3–4 m occurred along a more than 250 km long segment of the fault, with a maximum slip of 7 m. The rupture also reached the city of Antakya, which we mentioned in connection with the tragic earthquakes of 115 and 526.

The Mw 7.8 earthquake and subsequent aftershocks undoubtedly led to the initiation of the Mw 7.6 earthquake on the Çardak Fault. The epicentre of this second earthquake was approximately 90 km north of the East Anatolian Fault. The earthquake originated at a depth of approximately 7 km. From the hypocentre, the rupture propagated towards the East Anatolian Fault at a velocity of approximately 2.8 km/s, which is less than 90% of the S-wave speed in the area. Towards the east, however, the rupture propagated at approximately 4.8 km/s, which is supershear velocity. The total length of the ruptured portion of the fault was approximately 170 km. The rupture reached the East Anatolian Fault. Slip greater than 6 m occurred along a more than 60 km long section of the fault, with a maximum slip of 7 m.

The recorded values of seismic acceleration are also interesting. At several locations, the acceleration reached values greater than 1 g. The maximum acceleration value, 1.62 g, was recorded at the seismic station in the town of Fevzipaşa in the Gaziantep province. These values suggest a more complex fault geometry and consequently a relatively complex rupture propagation.

Several values will likely be further refined. Also, the initial indications of supershear propagation will need to be analysed in detail.

Although the earthquakes and aftershocks were relatively well instrumentally recorded, their analysis is not simple. This is due to the material and geometric complexity of the Earth's interior, the complex spatial distribution of stress and deformation, and finally, the size of the affected area.

We can say that the occurrence of such an earthquake doublet—Mw 7.8 and Mw 7.6 within 9 h—and in such proximity is literally a unique event.

Earthquakes urgently highlight the problem of estimating or determining the accumulation of stress and strain after a previous comparable earthquake in the area. The earthquakes have undoubtedly changed the state of stress and strain throughout the Anatolian Plate region. This change is a nontrivial challenge for seismologists. It is necessary to update the earthquake hazard analysis of the area of Turkey and the immediate surroundings.

Was the February Series of Earthquakes a Surprise?

Turkish, as well as other seismologists, were well aware that earthquakes occurred on the East Anatolian Fault. Significant earthquakes (according to available sources) include the following, listed in sequence according to their location on the East Anatolian Fault from its northeast end to its southwest end: 1866 (M 7.2), 1971 (M 6.7), 1789 (M 7.2), 2010 (M 6.1), 1874 (M 7.1), 1875 (M 6.7), 1905 (M 6.8), Sivrice 2020 (M 6.8), 1104 (M 6.5),

Pazarcik 1893 (M 7.1), Marash (Kahramanmaraş) 1114 (M 7.4), Marash (Kahramanmaraş) 1513 (M 7.4), Kahramanmaraş 1795 (M 7.0), 1822 (Mw 7.5), Amik (Antiochia, Antakya) 1872 (M 7.2).

These data indicate that relatively large earthquakes have repeatedly occurred in the Kahramanmaraş area over the past 3–4 centuries. It is not easy to locate or quantify the ruptured parts of faults in the case of historical earthquakes. Sometimes there are not even historical documents that would allow such a thing. Estimating the magnitudes of historic earthquakes is also difficult and sometimes nearly impossible. Certainly, such knowledge would allow seismologists to more accurately identify long-term quiet seismic fault segments and the amount of deformation that may have accumulated on a given fault segment since the last earthquake. This would still require the best possible knowledge of the geometry and rheology of the faults and the velocity and geometry of relative plate motion.

As we have mentioned, at the northeast end of the East Anatolian Fault, in the Karliova area, the velocity of relative plate motion is approximately 1 cm/year, around Türkoğlu it is approximately 4 mm/year, and at the southwest end, it is only about 2.5 mm/year on the main fault and approximately 1 mm/year on secondary faults.

Due to the geometry and direction of plate motion and significant variation in motion velocity along the fault, the East Anatolian Fault exhibits significant geometric complexities, manifested in heterogeneous seismic activity with quiet zones, earthquake clusters, and diffuse activity.

Some authors believe that the February earthquakes occurred in a quiet zone where an earthquake occurred in 1114.

Since the February earthquakes occurred, it is evident that the entire area was very close to a critical state of stress before 6 February 2023. It was enough for the strength limit of the fault contacts (determined by static friction) to be reached in some location, leading to the formation of a rupture whose propagation caused stress variation in the entire area "prepared" for an earthquake.

If seismologists were not aware of this state of approaching critical stress, it means that some of the data is not accurate enough or is simply missing. A simple example is that the magnitudes of previous earthquakes in the area are overestimated. As a result, it is assumed that so much strain and stress was released in those earthquakes that, at the known rates of plate motions, sufficient strain and stress could not yet have accumulated in a given segment.

Note that, in general, it is not rare that an area was near a critical stress state. There are many areas in which an earthquake will "easily" occur.

Why Were the Earthquakes So Tragic and Devastating?

The death toll and damage to the buildings were considerably greater than they could have been if the buildings had been of better quality. The tragic lesson of the 1999 Izmit earthquake has been sadly overlooked here. In rebuilding the devastated landscape, it will be necessary to reassess the level of seismic hazard, building standards and develop a quality assurance system for construction. Without this, Turkey would be facing another unnecessary tragedy.

11

An Optimistic Ending: Seismometers on the Moon and Mars

The previous chapters have shown that the major earthquakes have indeed been tragic challenges to humanity. The devastating events of Lisbon, San Francisco, Chile, Mexico, Indonesia, Japan and Turkey brought destruction and tragedy, but also new insights into the processes within the Earth. They have often led to important decisions in favour of protecting people and buildings.

As mentioned in the chapter on the basics of seismology, thanks to the unique properties of seismic waves, earthquakes have made it possible to learn about the interior of our planet. The seismic model of the Earth is the most accurate model of its interior that can be achieved.

Seismologists have realized that seismic waves are a universal tool for learning about the internal structure of the solid bodies of the Solar System. That's why they placed seismometers on the Moon and Mars. The seismometers will be among the first instruments transported to other similar celestial bodies. This is because they will make it possible to find out whether the bodies may be subject to quakes that could pose a threat, to find mineral deposits and, last but not least, to study the internal structure of the bodies. The knowledge of internal structures and processes in the bodies of the Solar System is very important for research into the evolution of planets and, consequently, of our Earth.

P. Moczo et al., *Earthquakes*, Springer Praxis Books,
https://doi.org/10.1007/978-3-031-64707-9_11

Moon

Currently, most scientists agree on the theory that the Moon formed around 4.5 billion years ago. A large celestial body, roughly the size of Mars, collided with the young, molten Earth, and material ejected from this impact into orbit around the Earth eventually coalesced, forming the Moon.

The Moon in the Ancient Human History

What role has the Moon played and continues to play for humanity on Earth? Alongside Venus, the Moon is the only celestial body that can be seen even after the sunrise. In the vicinity of the Earth's poles, where the Sun does not show its friendly face for long months during polar nights, the Moon is a more dominant celestial phenomenon than the Sun itself.

Human communities have always observed the events in the sky. In the Stone Age, they began to observe the Moon and the change in its shape, which regularly repeated. They found that this cycle lasts 29.5 days. At the same time, they noticed that while all the seasons change, the cycle repeats 12 to 13 times. Each full moon observed within one cycle received its own name. And so humans invented the lunar calendar.

The oldest preserved documentation of the Moon dates back to prehistoric times. Between 35,000 and 30,000 years BC, people recorded what we consider lunar calendars on bone fragments. Scientists have even proposed a theory that a needle with 29 notches, found in the South African Lembobo Mountains, could have served prehistoric women as a tool for tracking their menstrual cycles. Perhaps the most famous astronomical artefact is the so-called Nebra Sky Disk from the Bronze Age, which was found in Germany in 1999. It is the oldest known object depicting specific celestial bodies and the Moon in a stylized crescent form.

The Moon became an indispensable measure of time and played a significant role in shaping traditional beliefs and religious practices. In ancient Greco-Roman religions, the Moon was associated with the goddess of nature, vegetation, and childbirth Artemis/Diana, the embodiment of the Moon, Selene/Luna, or the goddess of crossroads, ghosts, magic, and protector of witches Hecate.

As they gazed upon the night sky, the Greeks came up with many explanations for the presence and character of the Moon, many of which seem bizarre today. However, in the fifth century BC, the philosopher Anaxagoras of Clazomenae (c. 500–428 BC) rejected the supernatural and divine nature of the Moon and the Sun. According to him, both the Sun and the Moon were actually large round rocks, and the Moon reflected light emitted by the heated Sun. These unorthodox ideas, stemming from what was then unusually scientific astronomical observation, contributed to the imprisonment and ostracism of Anaxagoras.

Anaxagoras attempted to explain the phases of the Moon and its eclipses. Based on his observations, he deduced several facts that were not so obvious at the time: if the Moon shines in the sky during the night, logically, the Sun must also be present at night. Until then, the prevailing belief was that a new Sun rose in the sky every morning, personified in the god Helios riding a golden chariot pulled by winged horses. If the Moon reflects sunlight and, at the same time, undergoes regular cycles of changing shape or even causes solar eclipses, it must be closer to the Earth than the Sun.

Aristarchus of Samos (c. 310–230 BC) determined the size of the Moon's radius during its eclipses from the ratios of the Earth's shadow and the Moon to be approximately 1/3 of the Earth's radius, which is quite close to the current value (0.27). Later, he also determined the relative sizes of the Sun, Earth, and Moon, as well as their mutual distances. Although his estimates were not quantitatively accurate, he deduced that the Sun was the largest. Aristarchus also considered a heliocentric model as an alternative to the geocentric model.

Finally, his idea was repeatedly confirmed—the same has been preserved in the works of Chinese philosophers from the Han Dynasty (206 BC–220 AD), as well as in the work of the Indian mathematician and astronomer Aryabhata (476–550 AD). Even Greek philosophers such as Parmenides (c. 540–470 BC) and Empedocles (c. 490–430 BC) advocated the theory that the Moon is not a source of light but reflects sunlight. According to Parmenides, the Moon was "a lamp in the night, wandering around the Earth with borrowed light." About 600 years later, Plutarch (c. 44/46–119/125 AD) wrote that the Moon is covered with countless irregularities, fissures, and rivers that cast shadows when sunlight falls upon them.

During the Crusades in the Middle Ages, the symbol of the crescent moon became closely associated with the Orient, through the goddess Hecate, who had an important place in the ancient Greek colony of Byzantium. Byzantium later became the Eastern Roman capital Constantinople, and it retained the symbol of the crescent moon. By the eighteenth century, the crescent moon symbol definitively entered Ottoman symbolism. In 1793, it was even decreed that the crescent moon symbol should mark every Ottoman vessel. The crescent moon is still depicted on flags of predominantly Islamic states, such as Turkey, Algeria, Pakistan, Malaysia, and many others.

However, the crescent moon is also present in Christian tradition. Since the Middle Ages, it has often accompanied depictions of the Virgin Mary. Evidence of the enduring enthusiasm for the celestial spectacle can be found in artistic expressions ranging from medieval illuminations to works by artists such as Giotto, da Vinci, and romantic depictions of European panoramas.

The Moon in the Modern Times

Renaissance depictions of the Moon show a sophisticated knowledge of its relationship to the Earth and the Sun. By the Middle Ages, the spherical nature of the Moon was already generally accepted, but despite Plutarch's opinion, the belief that it was perfectly smooth and reflected the surface of the Earth prevailed. The Renaissance scholar and artist Leonardo da Vinci (1452–1519) believed the Moon itself was dotted with various spots. This is evident in his texts and sketches.

The Moon played a very important role in the fifteenth and sixteenth centuries. The broadening of astronomical knowledge was essential for overseas voyages and the discovery of new worlds. When Christopher Columbus (1451–1506) and his crew were shipwrecked in Jamaica on their fourth voyage, the population had already had unpleasant experiences with Europeans. Christopher Columbus gained the respect of the locals and especially their willingness to share their food by predicting a lunar eclipse. Allegedly, he used the almanac of the astronomer Regiomontanus (1436–1476) from the Universitas Istropolitana in Pressburg (now Bratislava in Slovakia).

In November 1609, the Italian astronomer and scholar Galileo Galilei (1564–1642) recorded the results of his observations of the Moon with a self-made telescope from his home in Padua. He demonstrated the revolutionary level of his knowledge of the universe with his work *Sidereus Nuncius* (*The Starry Messenger*). His observations were highly acclaimed, and he was invited to Rome with great honour after papal astronomers from the Collegium Romanum independently confirmed Galileo's observations of 1609/10. The controversy came much later when Galileo interpreted his observations in heliocentric terms in his famous work *Dialogo sopra i due massimi sistemi del mondo* (*Dialogue Concerning the Two Chief World Systems*). However, he could not provide evidence at the time. The necessary evidence was not found until more than 100 years after Galileo. Galileo's heliocentric doctrine was branded heretical at his trial in 1633, contradicting the knowledge and understanding of the Holy Scriptures at the time. In any case, Galileo was intuitively right. Pope John Paul II acknowledged the error on the part of the Church and rehabilitated Galileo in 1992.

In 1609, a few months before Galileo's observations, the English mathematician Thomas Harriot (1560–1621), financially supported by the Earl of Northumberland, purchased a telescope in the Netherlands. He was thus 4 months ahead of Galileo in observing the Moon with a telescope. One of the first maps of the Moon was made by his hand. In 1647, the first work devoted exclusively to the Moon, *Selenographia* by the Polish-Lithuanian astronomer Johannes Hevelius (1611–1687), was published. The work gained great respect in Protestant countries, where it was used until the eighteenth century. In religiously divided Europe, an alternative work by the Jesuit astronomer Giovanni Battista Riccioli (1598–1671) quickly emerged. Riccioli named craters on the surface of the Moon after famous philosophers and astronomers. Giovanni Battista Riccioli and Francesco Maria Grimaldi (1618–1663) were the first to refer to the dark patches visible on the surface of the Moon as „maria "(Latin for seas) and the bright parts as „terrae "(Latin for continents).

In 1753, Croatian astronomer and philosopher Rudjer Josip Bošković (1711–1787) discovered that the Moon is not surrounded by an atmosphere. Much later, it was shown that lunar craters are the result of asteroid and meteoroid impacts.

USA and USSR in the Competition for the Moon

On 29 July 1958, President Dwight D. Eisenhower (in office from 1953 to 1961) signed the National Aeronautics and Space Act, which established the National Aeronautics and Space Administration (NASA). NASA is an independent agency of the US federal government responsible for the civil space program, aeronautics research, and space exploration.

The Soviet Luna space programme in 1959 can be considered the beginning of efforts to explore the Moon physically. The tiny Luna 1, with its five antennas, was the first spacecraft to get close to the Moon. It was soon followed by Luna 2, which shocked the world as the first man-made object to hit the lunar surface. Luna 3 provided the first grainy photograph of the Moon's far side.

On 12 April 1961, the Vostok 1 spacecraft carried Soviet cosmonaut Yuri Alexeyevich Gagarin (1934–1968) into orbit around the Earth. Gagarin was thus the first man in space.

The US public was very sensitive towards the Soviet primacy. That probably contributed to US President John Fitzgerald Kennedy (in office from 1961 to 1963) asking the US Congress on 25 May 1961, to commit the federal government to a program aimed at landing a man on the Moon by the end of the 1960s. Thus began a very ambitious NASA program called Apollo.

The goals of the Apollo project were more ambitious than just landing on the Moon and returning safely to Earth. The Apollo program was aimed at creating the technology to meet other US interests in space, achieving US primacy in space, conducting a program of scientific exploration of the Moon, and developing human capabilities to operate in a lunar environment. The mission's science program included the installation of a television camera, equipment to measure the composition of the solar wind, a seismic station, a laser remote retroreflector, and the collection of rock samples for transport back to Earth.

In the first phase of the Apollo programme, NASA sent several probes to collect all the information necessary to meet the main goal of the programme—a successful landing of man on the Moon. As part of the research, NASA was able to obtain photographs of up to 99% of the lunar surface.

In 1966, the Soviet Luna 9 was the first to land safely on the lunar surface and took the first photographs of the lunar panorama. Three months later, Luna 10 became the first artificial orbiter of the Moon.

In December 1968, the Apollo 9 crew was the first to leave Earth orbit and successfully orbit the Moon. Finally, on 20 July 1969, mankind's long-held dream came true: man reached the Moon. The crew of Apollo 11 also successfully achieved the national goal set by President John F. Kennedy: to make a manned lunar landing and return to Earth. The Apollo 11 crew consisted of American astronauts Neil Armstrong (1930–2012), Edwin E. Aldrin Jr. (later known as Buzz Aldrin, born 1930) and Michael Collins (1930–2021). The first two exited the Eagle spacecraft and entered the lunar Sea of Tranquillity (Mare Tranquillitatis).

NASA estimates that 650 million people worldwide watched the landing on the Moon. This was an unprecedented achievement for the mankind. Samples of material that reached US laboratories thanks to the Apollo missions were donated by the US government to several countries as diplomatic gifts.

By the time it ended in 1972, the Apollo programme had nine successful missions. In the 1970s, both Soviet and American spacecrafts brought back to Earth samples that contributed significantly to knowledge of the Moon's formation and its mineral composition.

Seismometers on the Moon

Without seismometers on the Moon, seismologists could only theorize about moonquakes and the propagation of seismic waves inside the Moon. Planetary seismology could have begun as early as 1959 as part of the Ranger program. Three Ranger probes were supposed to deliver seismometers developed at the California Institute of Technology (Caltech) to the Moon. However, the missions were not successful.

The true beginning of planetary seismology was the Apollo 11 mission. Buzz Aldrin successfully installed the first seismic station, which contained two extremely sensitive seismometers (Fig. 11.1). One was a three-axis long-period seismometer with a natural period of 15 s. (A long-period seismometer is the most sensitive to motions with long periods.) The other was a short-period seismometer with a natural period of 1 s. (A short-period seismometer is the most sensitive to motions with short periods.) The seismic station survived the first night but mainly due to problems with temperature stopped operating at noon after the first night.

Fig. 11.1 Edwin Eugene "Buzz" Aldrin Jr. installs a seismometer on the Moon. NASA public domain

If any seismic event (such as a moonquake or meteoroid impact) is to be detected on the seismometer record, it must be distinguishable from seismic noise (any disturbing background). Since the Moon lacks both an atmosphere and oceans, seismic noise on the Moon is significantly lower than on Earth. In fact the noise was not really resolved and was likely below the resolution of the Apollo seismometers.

Until April 1972, crews from the Apollo 12, 14, 15, and 16 missions installed four sensitive seismic stations. This created a network of stations that enabled the localization of seismic events. A radioactive thermoelectric generator ensured their long-term continuous operation. All seismometers, except for the short-period seismometer installed by the Apollo 12 mission and the vertical component of the long-period seismometer installed by the Apollo 14 mission, were in continuous operation until the end of September 1977, when their activities were terminated at Earth's command. Thus, the network recorded seismic motions on the surface of the Moon for almost 8 years. More than 600 gigabits of seismic data were sent to Earth.

The Apollo missions also placed geophones on the Moon, which were used in active seismic experiments. Geophones were capable of accurately recording seismic motion at frequencies ranging from 8 to 100 Hz.

The Apollo 17 mission installed a gravimeter on the Moon in December 1972. The primary goal was the detection of normal modes of the Moon excited by gravitational waves as suggested by astrophysicist Joseph Weber. A design error made that instrument unable to operate at its sensitivity.

Within the Apollo program, three types of seismic experiments were conducted, which we will explain on the following pages: passive seismic experiments at the landing sites of Apollo 11, 12, 14, 15, and 16 missions; active seismic experiments at the landing sites of Apollo 14 and 16 missions; lunar seismic profiling in the landing area of the Apollo 17 mission (Fig. 11.2).

Seismic records within the Apollo program were analysed promptly. However, significantly more powerful computers and new methods for analysing seismic records now allow seismologists to revisit Apollo seismic records and reanalyse not only records of seismic events but also records of continuous seismic noise. This enables them to gain new and complementary insights into the structure and seismic activity of the Moon.

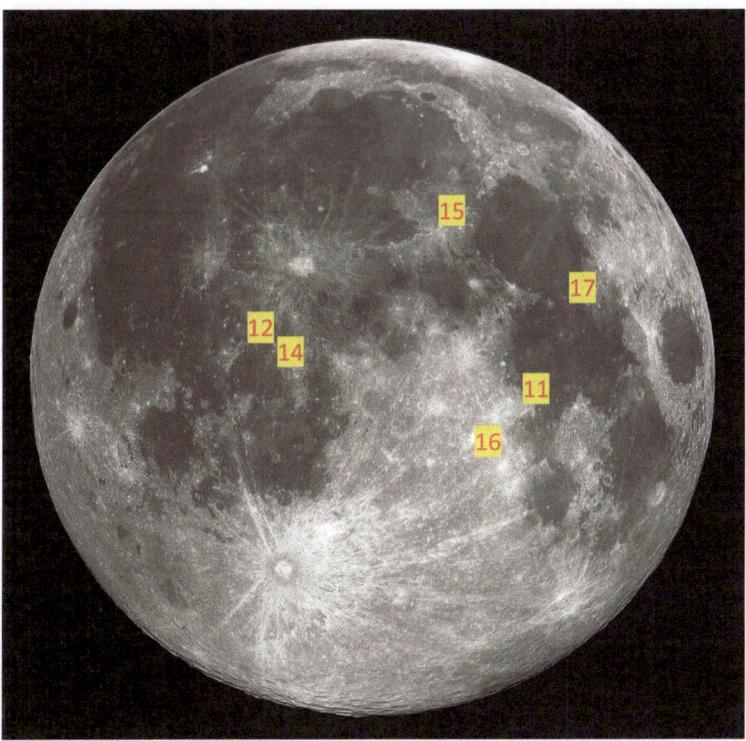

Fig. 11.2 Landing sites of the Apollo program missions on the Moon. At each site, at least one seismometer was used to record moonquakes, meteoroid impacts, artificial explosions, and controlled impacts of parts of rockets and lunar modules. Reprinted with permission from Moczo et al. (2023), © GRADA Slovakia s.r.o., 2023. All rights reserved

Events on the Moon in a Passive Seismic Experiment

Records of the seismic stations allowed seismologists to identify and analyse six types of seismic events.

Artificial Impacts of Objects on the Lunar Surface During the Apollo project, nine controlled impacts of the last stages of the Saturn V rocket (before the astronauts landed on the Moon) and the ascending parts of the lunar modules (after the crew returned from the Moon to the command and service module) were made on the lunar surface. The remaining fuel in the module was used for the controlled impact. The controlled impact was not only powerful, but also a very useful surface seismic source, as it allowed the time and location of impact to be known. In the case of the Apollo 12 mission, the impact was equivalent to about one ton of TNT in terms of seismic waves generated.

The recorded signal was a big surprise for seismologists. It was unlike anything recorded on Earth. It had a gradual increase in amplitude at the beginning, large amplitudes lasting for more than half an hour and then very slowly decreasing to noise levels for more than an hour. Such a signal is due to the extremely low attenuation and intense scattering of seismic waves inside the Moon—we will explain later.

The Impacts of Meteoroids The Apollo seismometers recorded more than 1700 meteoroid impacts, ranging from small particles to large boulders. Although the usefulness of the records is limited, as we only know about the impact from the records and do not know directly the time and location of the impact, they provide important information when combined with records of other seismic events.

Meteoroid impacts also provide information about small interplanetary objects crossing the Earth-Moon system. A network of seismic stations made the entire Moon a detector of these meteoroids. Analysis of the seismic records led to the detection of two distinct types—cometary (mostly with masses less than or approximately equal to 1 kg) and asteroidal (mostly with masses greater than 1 kg).

Deep Moonquakes The most common seismic events recorded are deep moonquakes, which occur at depths of 700–1200 km. In absolute comparison, at depths greater than those at which the Earth's deepest earthquakes occur. These moonquakes were a big surprise. Recall the radii of the Moon and the Earth: 1738 km and 6378 km.

During the Apollo project, more than 1300 deep moonquakes were identified, but later analysis identified more than 7000.

More than 300 clusters of moonquake sources have been detected so far. Locating the sources is complicated by strong scattering in the layer of broken rocks, which creates a long 'train' of waves on the seismic record (coda waves). Each cluster of sources generates a specific characteristic seismic record, which helps to better understand the wave content of the records. The spatial distribution of epicentres does not appear to be random and is related to the spatial distribution of the basaltic lunar maria.

The deep moonquakes in a given cluster occur with monthly, 7-monthly and 6-yearly periods, corresponding to the Moon's orbital motion. This suggests that the deep moonquakes are triggered by release of tidal stresses and background stress. The mechanism is not yet known. Some characteristics of

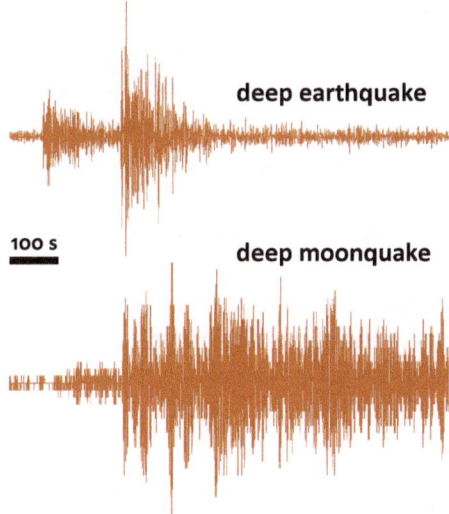

deep earthquake

100 s

deep moonquake

Fig. 11.3 Comparison of seismometric records of deep earthquake and deep moonquake

deep moonquakes are reminiscent of earthquakes that occur at depths of 70–300 km below the Earth's surface (Fig. 11.3).

Shallow Moonquakes Significantly fewer, only 28, shallow moonquakes with hypocentres at depths of up to 300 km were identified. They were originally referred to as "high-frequency teleseisms" because they contained waves with large amplitudes at high frequencies.

The depth of the hypocentres is difficult to estimate as all identified shallow moonquakes originated outside the network of the four seismic stations. Those moonquakes whose epicentres are located inside the area of the station network can be located with considerably more precision.

Shallow moonquakes have a significantly larger fraction of high frequencies, even though they originated at large distances from seismic stations, and one would expect high frequencies to be sufficiently attenuated at such large distances. The large epicentral distance was better reflected by the temporally separated arrivals of the P and S-waves.

Shallow moonquakes are the largest seismic events apart from the impacts of large meteoroids. The largest detected moment magnitude was 4.1.

The low number of recorded shallow moonquakes does not even allow statistical analysis. No correlation with tidal forces was found. The available data do not yet allow a plausible determination of the causes and mechanism of shallow moonquakes.

Thermal Moonquakes A number of weak moonquakes can also be identified in the seismic records of short-period seismometers and geophones, which, given the timing and recurrence of their occurrence, are likely to release stresses caused by abrupt temperature changes during sunrise and sunset. At these times, the temperature varies approximately between -170 °C and 127 °C. Landslide of regolith (unconsolidated material on the lunar surface) on the slopes is thought to be a possible main mechanism.

Astronaut Activities On the seismic records, seismologists could also well identify the astronauts' walking on the lunar surface, ascents and descents of the lunar module stairs, vibrations caused by the lunar rover and other specific activities.

Unclassified Seismic Events Finally, one can also speak of unclassified seismic events in cases where the weak signal on the seismic record has not yet been attributed to any identifiable source.

The catalogue of seismic events published in October 2008 includes 9 controlled impacts of objects, 1744 meteoroid impacts, 28 shallow moonquakes, 7400 deep moonquakes, over 1000 thermal moonquakes, and 3322 unclassified seismic events. In total, over 13,000 seismic events.

It is estimated that Earth releases about 10^9 times more energy in the form of seismic waves than the Moon. This is because Earth has lithospheric plates in motion, which create tectonic activity. The Moon lacks such tectonics because it is smaller and cooler than Earth and does not have mantle convection.

Another important difference is that moonquakes exhibit a certain periodicity related to orbital periods, while the occurrence of earthquakes in time is random.

The seismometric records on the Moon significantly differ from records of earthquakes or explosions on Earth. In most cases, the recording gradually begins without a distinct onset, amplitudes rise to a maximum, and then decrease very slowly, often over hours. There are no visible clear arrivals (phases) within the record. Such a record is the result of extremely low attenuation of seismic waves (dissipation of vibrational motion energy) and intense wave scattering near the surface. The low attenuation can be explained by the absence of water, at least in the near-surface regions of the Moon. Extensive scattering is a consequence of the Moon's surface being pulverized by numerous asteroid impacts. The lunar crust may have such characteristics to a depth of tens of kilometres.

This environment is so unique that the propagation of vibrational motion in it can be roughly described as a diffusion process. This applies to waves generated by impacts at large distances as well as waves generated by lunar vehicles at small distances.

Active Seismic Experiments

Active experiments were conducted as part of the Apollo 14 mission in the Fra Mauro region of Mare Procellarum and the Apollo 16 mission in the southeastern highlands region of the Moon. The active seismic experiment consisted of a 91 m long profile of three evenly spaced geophones, explosive sources applied at 4.57 m intervals along the geophone profile, and also mortar launchers. However, due to concerns about potential damage to the equipment, mortar launchers were not used at the Apollo 14 site, and those at the Apollo 16 site were used 450 m away from the furthest geophone.

The results from the active experiments showed that these sites were covered by a layer with a thickness of 8.5 m (Apollo 14) and 12.2 m (Apollo 16) with P-wave speeds of 104 m/s and 114 m/s, respectively, indicating porosity or fragmented rock. Beneath this surface layer is a layer with a thickness of at least 1 km and P-wave speeds of 299 m/s (Apollo 14) and 250 m/s (Apollo 16). These speed values also indicate fragmented rock.

Lunar Seismic Profiling Experiment

The Lunar Seismic Profiling Experiment was conducted during the Apollo 17 mission at the Taurus-Littrow site on the southeastern edge of Mare Serenitatis. Four geophones were placed in a T-shaped configuration with the greatest distance approximately 100 m between geophones. Astronauts prepared eight explosive sources placed at distances up to 2.7 km from the geophones. These explosives were detonated only after the astronauts had left the Moon. The impact of the ascent stage of the lunar module was also recorded by the geophones at a distance of approximately 9 km from them.

In addition to the surface layer with a P-wave speed of 100 m/s, corresponding to lunar regolith, layers with speeds of 327 m/s, 495 m/s, and 960 m/s were found. The layer with a speed of 960 m/s extends to a depth of at least 3 km or deeper.

Interior Structure of the Moon

Despite all efforts and the abundance of seismic records, the data obtained so far remain incomplete and challenging to interpret. The fundamental limitation lies in the fact that the installed seismometers covered a relatively small portion of the Moon's total surface (Fig. 11.2).

The latest analyses of seismic records significantly differ from those conducted during the Apollo program. This difference is partly due to new methods of signal analysis and the incomparable computing power available then and now. Methodological advancements can also be illustrated by the fact that records of seismic noise, which were unusable back then, are now an indispensable source of new information.

Currently, it is highly likely that seismic data from the Apollo missions allow for estimating the structure of the upper and middle parts of the mantle to a depth of approximately 1200 km on the Moon.

Even under these circumstances, it has been possible to develop a certain image of the Moon's interior. The Moon is composed of three main geochemically distinct layers—the crust, mantle, and core. Each of these three layers consists of several sublayers. The thicknesses of the layers are only estimated and require further investigation (Fig. 11.4).

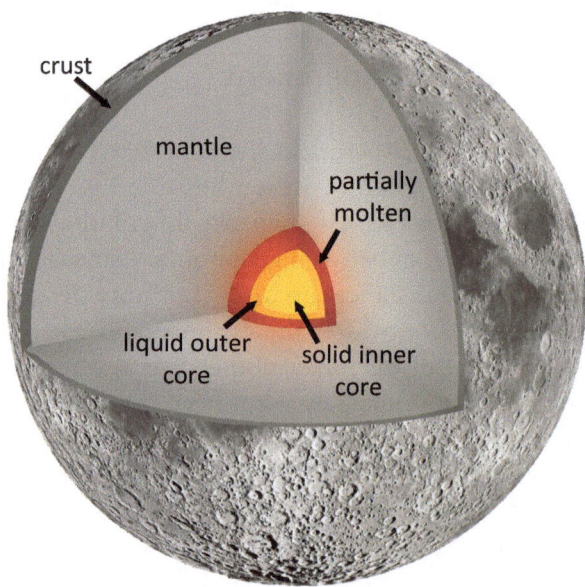

Fig. 11.4 A simplified estimated model of the Moon's interior developed primarily based on the interpretation of recorded seismic waves

Seismologically, the lunar crust can be divided into a thin surface layer that strongly scatters seismic waves and a lower part where pressure acts to smooth the rocks thus reducing the degree of scattering.

Beneath the crust lies the mantle, which may be divided into an upper mantle and a lower mantle approximately 500 km below the surface of the Moon. Whether this material boundary exists throughout the Moon and whether it is equally prominent everywhere will be the subject of further research. The attenuation of S-waves suggests that the lowest part of the mantle is partially molten. Seismic speeds throughout the mantle are not well determined.

In this paragraph, we'll provide several numerical values, primarily for readers interested in comparing them with values inside Earth and Mars. Other readers can skip this paragraph without any issue. Approximate values of P-wave speeds, S-wave speeds, and densities at depths down to 100 km are 1.0–7.8 km/s, 0.5 – 4.4 km/s and 2600–3360 kg/m^3. Approximate values of P-wave speeds, S-wave speeds, and densities at depths of 100–1200 km are 7.6–8.3 km/s, 4.4–4.7 km/s and 3340–3440 kg/m^3.

There are certain indications of a zone of decreased speeds in the upper mantle at depths of approximately 100–250 km, which is consistent with a temperature gradient of around 1.7 °C/km. This temperature gradient could be related to the presence of a thermally anomalous area known as the Procellarum KREEP Terrane, which contains large amounts of radioactive elements producing heat.

The Moon has a small liquid core and a solid inner core. To determine their thicknesses, a reliable mantle model would be necessary. The liquidity of the outer core and the existence of the inner core are inferred by analysing seismic waves that pass through the outer core as P-waves, reflect off the inner core and propagate as S-waves in the mantle.

The deepest interior of the Moon, as inferred from seismic data, poses a challenge to our understanding of its origin, evolution, and dynamic behaviour in several respects. The existence of molten rocks at great depths may be related to the Moon likely cooling through heat conduction, and the thermal conductivity of the surface unconsolidated layer is probably very low. Increased attenuation of S-waves may be caused by elevated water content. The question of water on the Moon is complex and still open. However, it seems that lunar rocks contain comparable or only 10 times less water than the Earth's mantle.

The observation of a solid inner core was completely unexpected and raises questions regarding the origin of planetary magnetic fields.

Mars

Mars is the fourth planet from the Sun and one of the five planets known in ancient times.

Mars was formed approximately 4.5 billion years ago. It formed from planetary embryos and rocky material from the inner part of a protoplanetary disk.

Mars has an average radius of 3390 km, has 10 times less mass compared to Earth, and a lower average density. These and other characteristics of Mars could be determined without instrumented landers on the Martian surface. However, finding out more about the Martian surface required bringing scientific equipment to Mars. To look inside Mars and learn about its structure was only possible by installing a seismic station. This has only recently been achieved.

But first, let's recall how people looked at Mars long before they were able to approach Mars and bring scientific instruments to Mars.

Millennia of Observing the Red Planet

The planet Mars was observed by the ancient Egyptians as early as the second millennium BC. The Egyptians documented Mars on astronomical maps in the tombs of pharaohs. The Sumerians also created similar maps, and knowledge of the planet Mars is documented in ancient China as well. These ancient civilizations associated life on Earth and phenomena in the heavens with their deities, and this knowledge, along with other understanding, was inherited by the ancient Greeks. They associated the planet Mars with the god of war, Ares.

The term "planet" itself comes from Greek, which in ancient times referred to the seven celestial bodies, according to Plato (428/423–348/347 BC), moving around the Earth in accordance with ancient Greek geocentric ideas—the Moon, the Sun, Venus, Mercury, Mars, Jupiter, and Saturn. Plato's student, Aristotle (384–322 BC), in the fourth century BC, observed the occultation of Mars by our Moon, which supported existing models of the arrangement of planets in our system. He believed that these objects orbit around the Earth at fixed distances and speeds.

The Greek mathematician and astronomer Claudius Ptolemy (c. 85/100–165/170 AD) in the second century AD pointed out that the speed of Mars' motion across the sky is not constant. At the same time, he proposed a new arrangement of the planets—closest to the Earth was the Moon, followed by Mercury, Venus, the Sun, Mars, Jupiter, and Saturn. Claudius Ptolemy wrote the work *Mathēmatikē Syntaxis* or *Hē Megalē Syntaxis* (later better known by its Arabic title Almagest) in Alexandria. This work encompassed all the astronomical knowledge of the time based on the geocentric model. It also included a star catalogue. This work remained an authoritative text on astronomy in both the European and Arabic worlds until the early modern period.

In the sixteenth century, Nicolaus Copernicus (1473–1543) explained with his heliocentric concept why Mars, Jupiter, and Saturn appeared on the opposite side of our sky compared to the Sun during its apparent retrograde motion.

At the beginning of the seventeenth century, Galileo Galilei utilized a telescope to peer into the cosmos and could, for the first time, observe Mars better than with the naked eye. However, the telescope had too low a resolution to reveal what lay on the surface of the planet. The first map of Mars, depicting polar ice caps and dark spots, was created by the Dutch astronomer Christiaan Huygens (1629–1695) in 1659. He estimated the planet's radius to be approximately 60% of Earth's radius. This estimation did not differ significantly from the current value of 53%.

The seventeenth century brought about new insights into Mars. Its volume was reassessed, and both the southern and northern polar caps were observed. The planet's rotation period was estimated to be 24 h and 40 min, which is only 3 min longer than the value known today.

Throughout the eighteenth century, Mars continued to attract the attention of astronomers, leading to increasingly precise and detailed information about the planet as a result of their observations. With the development of lenses and optics in the early nineteenth century, telescopes (both refracting and reflecting) improved, expanding the possibilities of their use. However, telescopes still could not capture observed images photographically. Astronomers spent hours behind telescopes, waiting for suitable conditions that would allow details of the observed object to be seen. They then sketched and recorded the planet at different times of the year, in different positions, and identified visible features on its surface.

Thanks to technological advancements, in 1840 German astronomers Wilhelm Beer (1797–1850) and Johann Heinrich Mädler (1794–1874) were able to compile the first detailed map of the surface of Mars.

Canals?

During the second half of the nineteenth century, several astronomers began to observe a previously unnoticed dense network on the surface of Mars, which evoked the idea of irrigation canals. Italian astronomer Giovanni Schiaparelli (1835–1910) observed Mars in 1877 and named them "canali" in Italian, referring to natural depressions. Later, in the English translation, popularized by an American entrepreneur Percival Lowell (1855–1916), the meaning shifted towards "canals", implying artificially constructed irrigation channels. However, the development of this theory is not as disconnected from reality as it might seem at the first glance.

The astronomers of that time believed that Mars was a dying planet. From their observations, it seemed that the natural processes on Mars were not so different from those on Earth—the vegetation changed colour and density according to the season, and the shape and extent of the ice caps also changed. The inhabitants of this planet supposedly attempted to save it from extinction by building an irrigation system to distribute water from the poles to the rest of the surface. What the astronomers of the nineteenth century did not know was that there was no vegetation on Mars and the changing colour spots were nothing more than dust storms and changes in surface albedo (surface reflectivity) due to seasonal CO_2 sublimation from the surface. The seductive but deceptive ideas about canals on Mars originated from the realities of the time of the astronomers themselves. They lived in an era of extensive canal constructions in the Suez and Panama canals.

The idea that Mars is an older planet than Earth typically led to the assumption that its inhabitants must be more advanced beings than those on Earth. There were even alleged sightings of moving lights interpreted as attempts by a Martian civilization to establish contact with life on Earth.

Gradually, a fantastic narrative emerged—an interpretation of how the intelligent inhabitants of a dying planet were making a final effort to save it and establish contact with their cosmic neighbours. These ideas were developed by the contemporary press, emerging science fiction literature, and "scientific" astronomical articles.

Mars on Earth

Scientists currently explore the possibilities of life on other planets in two ways—either using probes or telescopes, or inductively through the most precise simulations of conditions possible. They conduct these simulations in locations on Earth with the most extreme conditions for life.

In the nineteenth century, Percival Lowell focused on northern Arizona, where he studied the canyons, mountainous and desert conditions of the region, in an attempt to imagine what the landscape of Mars might look like.

Percival Lowell, convinced that Mars was a dying planet, was one of the advocates and popularisers of the theory that the inhabitants of Mars were trying to save their drying landscape by building an extensive irrigation system. In 1894, Lowell Observatory was founded in Flagstaff, Arizona.

Percival Lowell lived in a part of North America that suffered serious drought-related agricultural problems. In 1877, the Desert Land Act came into effect, aiming to irrigate and cultivate American lands afflicted by drought. Until the early twentieth century, both the government and private companies took measure after measure in an effort to cultivate barren land.

While Percival Lowell may have increased interest in Mars through his popularization efforts, Mars was certainly a subject of interest before him. French astronomer Camille Flammarion (1842–1925) was, like Lowell, a successful author of popular science literature on astronomy and science fiction works. He, too, believed that on the dying Mars, an advanced civilization was trying to cultivate what little remained of the planet. There were many astronomers of this kind in the nineteenth century. The public enjoyed their approachable writing style, entertaining lectures, and occasional indulgence in conspiracy and fantasy in discussion. However, this same behaviour often led them to be viewed as charlatans and popular authors rather than true scientists in the eyes of more professionally oriented colleagues.

As technology has advanced, astronomers have realized that the imperfect telescopes used in the past gave a blurry image. As a result, craters on the planet's surface were merging into indistinct lines. Through a confluence of three factors—the resolving power of the Grande Lunette refractor at Meudon, France, clear weather, and Mars in opposition—the Greek-French astronomer Eugène Antoniadi (1870–1944) succeeded in disproving the theory of Martian canals in 1909.

At the beginning of the twentieth century, theories that Mars' atmosphere is very similar to Earth's were debunked. In the 1920s, scientists succeeded in measuring day and night temperatures at both the equator and the poles of Mars. The temperature on Mars is much lower than astronomers initially thought, and the atmospheric pressure is too low for Mars to have areas of liquid water and a civilization similar to that on Earth.

Space Race for Mars

Scientific and technological progress allowed two superpowers to expand the battlefield of the Cold War beyond the borders of our Earth.

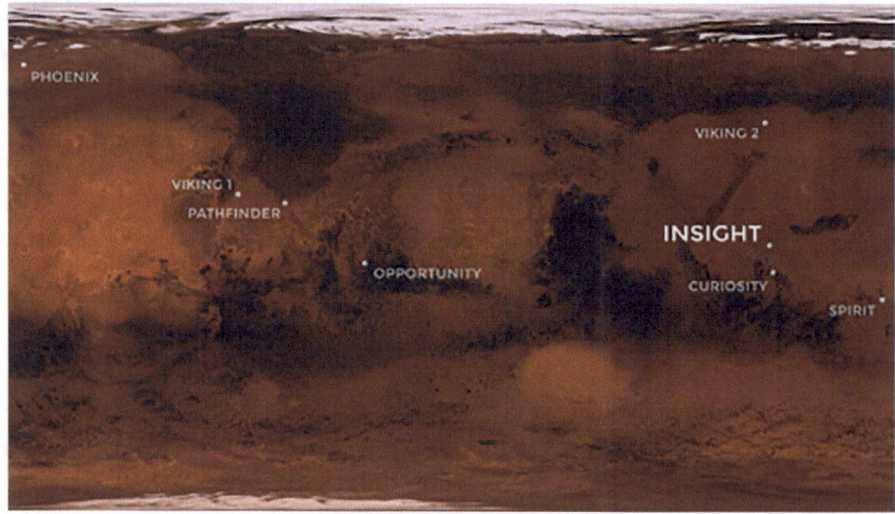

Fig. 11.5 NASA mission landing sites on Mars. The InSight mission has reliably detected marsquakes for the first time. NASA—public domain

The USSR attempted to reach Mars with a probe as early as October 1960, but the first 5 attempts failed. The first successful flyby of Mars was achieved by the USA in July 1965, followed by two more in 1969. The Soviet Mars 2 reached Martian orbit in November 1971, but the landing module with the vehicle crashed (hard landing) on the Martian surface. The first soft landing was achieved by the USSR with Mars 3 in December 1971. However, communication was lost 110 s after landing.

The USA achieved successful landings with two identical probes, Viking 1 and Viking 2 (Fig. 11.5). The Viking 1 module landed on Mars on 20 July, while the Viking 2 module landed on 3 September 1976. Among other instruments, the landing modules also included seismometers. However, the seismometer on the Viking 1 module failed to deploy and unlock properly. The seismometer on Viking 2 did deploy and unlock, allowing it to record seismic activity on the Martian surface. Despite nearly 19 months of nearly continuous operation, no credible marsquakes were recorded.

Twenty years later, the Russian Mars 96 mission (also referred to as Mars-8) was supposed to resume seismic measurements on Mars. However, the mission failed shortly after launch. We had to wait until 2018 for another opportunity.

Seismometers on Mars

The InSight Mission

On 5 May 2018, an Atlas V-401 rocket carrying the InSight robotic lander took off from Vandenberg Air Force Base in California. Built by Lockheed Martin Space Systems, InSight carried 8 unique scientific instruments. The main one was the very broadband seismometer SEIS (Seismic Experiment for Interior Structure). It was designed by the Planetary seismology team (led by Professor Philippe Lognonné) of the Institut de Physique du Globe de Paris, with colleagues at the French Space Agency CNES (coordinated by Dr. Philippe Laudet). The thermal probe (HP³—Heat Flow and Physical Properties Package) created by the German Aerospace Centre (Deutsches Zentrum für Luft- und Raumfahrt, DLR) was also important. InSight landed on Mars on 26 November 2018 at 11:52:59 AM Pacific Time in the Elysium Planitia region.

The InSight project (Interior Exploration using Seismic Investigations, Geodesy and Heat Transport) was part of the NASA Discovery program focused on the Solar System research. InSight was managed for NASA by the JPL (Jet Propulsion Laboratory) of the California Institute of Technology (Caltech).

The InSight project had two main research goals to significantly help understand the process that formed all the terrestrial planets in the inner Solar System:

- Research into the internal structure and processes inside Mars included determining the size and composition of the core (especially whether the core is liquid or solid), the thickness, structure and composition of the mantle, the thickness and structure of the crust, and the temperature and heat flow inside Mars.
- Research into the tectonic activity of Mars and meteoroid impacts involved the recording and analysis of seismic waves generated by marsquakes and meteoroid impacts.

Unique Seismometer

SEIS consists of a 3-axis very broadband seismometer, a 3-axis short-period seismometer, electronics and the deployment/protection system. The instrument was a collaboration between France, USA, UK, Switzerland, and

Germany. 3-axis means that seismic motion in three independent directions is recorded. A very broadband seismometer is capable of equally sensitively recording the particle velocity of seismic motion over a wide range of frequencies—from approximately 10^{-3} to 10^2 Hz. A short-period seismometer can do this in the band of about 1 to more than 10^3 Hz.

The SEIS seismometer is so sensitive that it can record the seismic motion of the Martian surface at the seismometer location with an amplitude smaller than a hydrogen atom!

The seismometer was placed on the surface of Mars on 19 December 2018 and has been fully operational since the end of February 2019. Achieving high-quality and reasonably analysable records of seismic waves generated by marsquakes and meteoroid impacts required an unprecedented technical effort.

On Earth, sensitive seismometers are often enclosed in special underground enclosures to isolate them as much as possible from changes in subsoil and air temperature, from changes in air humidity, and from nearby mechanical and electromagnetic interference sources.

The robotic module could not place the seismometer below the surface. Although the surface of Mars itself is relatively seismically quiet, the really big problem for a seismometer on the surface of Mars is temperature variation and strong winds.

Temperatures in the vicinity of InSight can range from nearly −100 °C at night to 0 °C during the day. Temperature changes on Mars can cause expansion and contraction of metal springs and other parts of the seismometer, as well as the cable connecting the seismometer to the lander. These changes can manifest themselves as noise in the seismometer recordings. Wind speeds can reach 100 km/h. Wind, especially during the northern winter season and more during the day than at night, causes unwanted seismic noise. Therefore, SEIS had to have sophisticated protection against these strong disturbances.

The first protection is the wind and thermal shield (Fig. 11.6). The aerodynamic shape of the shield prevents the seismometer from overturning in strong winds. The second protection is the construction of the seismometer itself. When some parts expand and contract due to temperature changes, others do so in the opposite direction to partially counteract these effects. The seismometer is vacuum-sealed in a titanium sphere, which is further protected by another insulation layer suitable for the specific thin atmosphere of Mars.

The cable connecting the seismometer to the module was later covered with Martian sand using the robotic arm. Insulating the cable reduced the effects of wind on it, thus also reducing potential noise in the seismic recordings.

An unexpected but significant technical complication arose from dust settling on the solar panels. The dust caused the panels to produce progressively

Fig. 11.6 The photograph from Mars shows the protective shield of the SEIS seismometer, which recorded seismic events on the surface of Mars during the period from February 2019 to December 2022 as part of the InSight mission. NASA—public domain

less electrical energy needed for the operation of the instruments. Due to weight constraints and the assumption that occasional strong winds would clear the panels of dust, InSight did not have a direct means of cleaning the solar panels. However, the cleansing winds did not come.

As long as the module had enough energy, the seismometer could reasonably record marsquakes and meteorite impacts (without significant masking noise) thanks to all the protective systems in place.

Seismic Activity of Mars

The InSight mission and the SEIS seismometer have indelibly made history in understanding the planets of the Solar System. For the first time, marsquakes were recorded on the Red Planet, providing evidence of its seismic activity.

In total, more than 1300 seismic events were recorded, with over 50 of them having clear enough signals for the InSight mission team to obtain

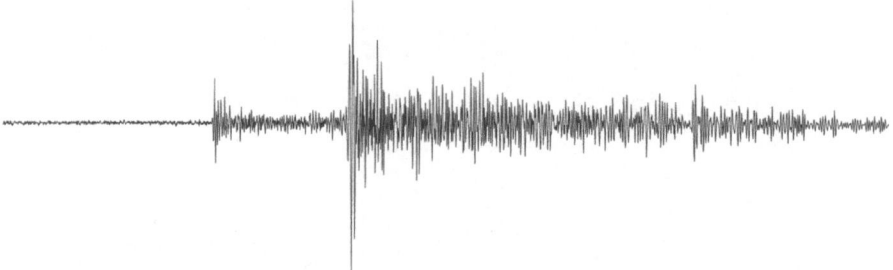

Fig. 11.7 The record of the marsquake on 25 July 2019 (time increasing from left to right). Clear onsets of two distinct wave groups correspond to P and S-waves. NASA—public domain

information about their locations on Mars (Fig. 11.7). The majority of well-recorded marsquakes occurred at depths of less than 40 km in the Cerberus Fossae region, approximately 1610 km away from the landing site.

Seismologists have identified two main categories of marsquakes. Records of high-frequency marsquakes have their maximum energy at frequencies above 2.4 Hz, which is the resonant frequency of the substrate at the seismometer's location. The recorded signal lasts 5–20 min, in most cases containing two distinct phases interpreted as P and S-waves in the crust. Records of low-frequency marsquakes last 10–20 min, in most cases exhibiting two distinct phases (likely corresponding to P and S-waves) and having their maximum energy at frequencies below 2.4 Hz, although some marsquakes also have energy above the 2.4 Hz frequency.

In the case of high-frequency marsquakes, the source is likely within the shallow part of the Martian crust. The low-frequency marsquakes likely originate at depths 30–40 km, that is, in the lower crust.

When the source is within the crust, P and S-waves primarily propagate within the crust. In the mantle, they are attenuated such that they cannot be distinguished from noise in the recordings.

Sources in the mantle are typically stronger. Due to their location, they do not generate sufficiently intense waves in the crust, but they generate waves in the mantle that can propagate over relatively large distances at low frequencies (long periods). At higher frequencies, seismic motion is damped.

Localization of earthquakes on Earth is better when more seismic stations record the earthquake. Certain knowledge of seismic wave speeds in the region is crucial. A seismic network would also be needed on Mars. However, since the InSight mission could only place one seismometer on Mars, seismologists, more than ever before, must utilize everything that the recordings from a single station allow.

Recording marsquakes from various distances is a crucial factor. The recordings differ in the types of waves they capture and the time differences between arrivals of different types of waves. When analysing marsquakes, it is particularly important to utilize all the knowledge gained from the analysis of previous marsquake events.

Seismic Model of Mars

The crust of Mars behaves like a mixture of Earth's and the Moon's crusts. Recall that the Moon's crust is considerably more fractured and cracked compared to Earth's crust due to asteroid impacts. Unlike Earth's crust, the Moon's crust lacks water (in liquid form) to fill the cracks with minerals. Therefore, seismic waves in the Moon's crust are significantly more scattered than in Earth's crust, arriving at the seismic station along many paths. Consequently, they arrive at different times, and duration of a moonquake recording is substantially longer than a recording of an earthquake of the same size.

Since seismologists only had seismic recordings from one seismic station, they could only determine the structure of the Martian crust directly beneath the InSight lander, and even that was not definitive. The results of the seismic wave analysis allow for two alternative interpretations. The Martian crust may consist of two or three layers. Derived seismic wave speeds are consistent with the idea of unconsolidated and heavily fractured material.

In the following three paragraphs, we will provide several numerical values, especially for readers interested in comparing the values inside the Earth and the Moon with those inside Mars. Other readers can easily skip these three paragraphs.

At InSight landing site the crust most likely consists of three layers. The depths of the interfaces between layers 1 and 2, and layers 2 and 3 could be approximately 4–23 km and 6–54 km, respectively, and the depth of the bottom of the crust could be 48–104 km. S-wave speeds could be approximately 2.2–2.9 km/s, 2.9–4.1 km/s, and 3.9–4.4 km/s. The ratios of the P-wave to S-wave speeds could be 1.73–1.77.

Seismologists utilized well-recorded low-frequency marsquakes with phases of seismic waves PP, PPP, SS, and SSS (Fig. 2.10) to determine the structure to a depth of approximately 800 km. The key finding is the thickness of the thermal lithosphere, estimated to 400–500 km. Seismologists estimated that the S-wave speed decreases from the crust, with a value of approximately 4.4 km/s, to about 4.25 km/s at a depth of approximately 500 km. From this depth, it increases to 4.4 km/s. The P-wave speed remains approximately

Fig. 11.8 A simplified estimated model of the interior of Mars developed primarily based on the interpretation of recorded seismic waves. NASA—public domain

constant up to a depth of 500 km, at around 7.8 km/s, and then increases to 8.2 km/s at a depth of 800 km.

The original analysis based on the detection of deep S-waves resulted in estimate of a core size of 1830 ± 40 km. However, subsequent analysis led to finding a liquid silicate layer between the mantle and the core which reduces the core size estimate to 1650 ± 20 km. The relatively large core indirectly indicates that the Martian mantle has a similar mineralogical composition to the mantle on Earth (Fig. 11.8).

No significant marsquakes were recorded from prominent volcanic regions, such as Tharsis, where three of the largest volcanoes on Mars are located. However, since Mars is not currently volcanically active (though Cerberus Fossae likely has some seismo-volcanic activity), this is not surprising. It is possible, though, that SEIS did not record those marsquakes that had hypocentres in the seismic shadow zone. For the given location of the seismometer, hypocentres within the shadow zone are those from which seismic waves in the mantle and crust cannot reach the seismometer's location because they would have to pass through the core. On Earth, direct waves can only be recorded up to a distance of 104 degrees. On Mars, the shadow zone caused

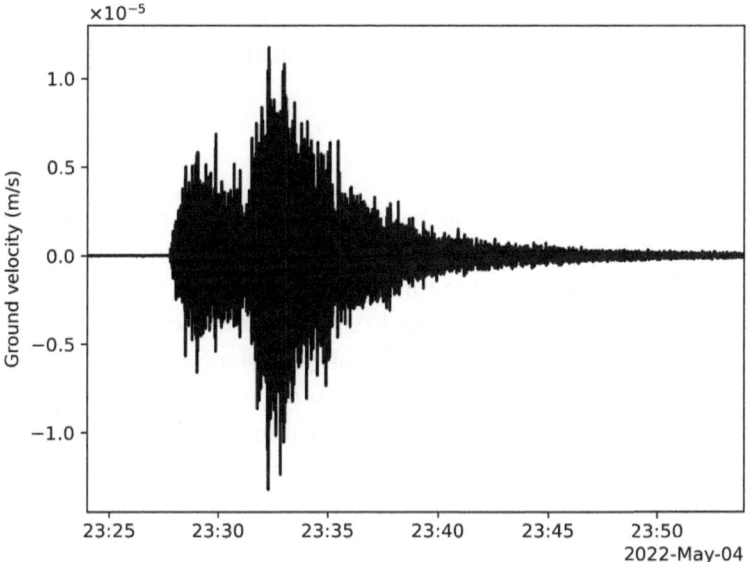

Fig. 11.9   On 4 May 2022, the SEIS seismometer recorded the largest marsquake with a magnitude of 4.7. NASA—public domain

by the relatively large core begins at a distance of approximately 94–98 degrees.

On 4 May 2022, SEIS recorded the largest marsquake with a magnitude of 4.7 (Fig. 11.9). When planning the InSight mission, seismologists hoped that SEIS would also record several such big marsquakes or even larger than magnitude of 5. Recall that on Earth, due to the dynamics of tectonic plates, there are approximately 170 earthquakes with magnitudes close to 5 in a year, on average.

Analysis of the seismograms revealed that most surface faults are not seismically active and that seismic activity mostly originates from a single region of tectonic structures, Cerberus Fossae. The character of deeper low-frequency marsquakes suggests a structurally weak, potentially warm source region, consistent with volcanic activity over the past two million years at depths of 30–50 km. High-frequency marsquakes occur along Cerberus Fossae, in the brittle shallow part of the crust.

Analyses lead to the conclusion that Cerberus Fossae represents a unique tectonic environment shaped by ongoing magmatic processes and locally elevated heat flow.

Orbital probes have also detected traces of boulders that may have rolled down steep slopes after being dislodged by marsquakes.

Meteorite Impacts

SEIS has also recorded seismic waves generated by nearly ten meteorite impacts. The impact locations of some have been confirmed thanks to flybys by NASA's Mars Reconnaissance Orbiter.

The question is why SEIS recorded relatively few meteorite impacts. Mars is located next to the main asteroid belt of the Solar System. The thickness of Mars' atmosphere is only 1% of the thickness of Earth's atmosphere. Therefore, more meteoroids pass through it without breaking up. It is possible that many of the impacts may have been obscured by wind noise or seasonal changes in the atmosphere.

However, since the characteristic seismic footprint of the impact on Mars has been found, scientists expect more impacts to be identified in the nearly 4-year SEIS records.

Analysis of meteorite impacts will be critical in refining the temporal evolution of Mars. Scientists can approximate the age of the planet's surface based on the number of craters caused by impacts. The more there are, the older the surface. By calibrating their statistical models based on how often they see recent impacts, they can estimate how many impacts have occurred in the history of the Solar System.

The seismic record of the impact can be used in combination with orbit images to reconstruct the meteoroid's trajectory and the size of its shock wave. Each meteoroid creates a shock wave in the atmosphere and an explosion when it hits the surface of Mars. These events generate sound waves through the atmosphere. The larger the explosion, the more distinct the seismic record.

New Era of Planetary Seismology and Physics of Terrestrial Planets

InSight also recorded invaluable data about the weather and magnetic field of Mars. Seismic data and other independent (geo)physical measurements allow for the determination of Mars' thermal history.

InSight represents a significant advancement in understanding the interiors of terrestrial planets. The knowledge gained can be applied not only to Earth, the Moon, Venus, but also to terrestrial planets in other planetary systems around distant stars.

InSight has opened a new era of planetary seismology and terrestrial planet physics and signifies an important step in the preparation of future missions.

Seismology on planets will continue, on the Moon with 3 seismometers to be deployed between 2026 and 2028 (FSS, LEMS for NASA, ChangE'7 for China) and with one seismometer deployed on Titan by the NASA mission DragonFly after 2030.

12

Earthquakes and Seismology of the Future

The stories of tragic earthquakes and tsunamis are not just interesting. All the events we have written about have been shocking surprises in many ways. However, they have ultimately led to significant advances in our understanding of what happens inside the planet we live on. And each time they also serve as a reminder of how complex the Earth's interior is, and how difficult it is to predict what will surprise us tomorrow, in a month or in a year.

What we do know is that there is no indication that earthquakes will become less frequent in the future. What does that mean for us?

Less developed countries with a moderate to high seismic hazard will continue to be most at risk, especially cities with high population densities and poor infrastructure. This situation is caused by rapid population growth due to migration from rural areas, lack of financial resources and corruption.

Even in developed countries, the vulnerability of cities is increasing due to the increase in population density and the technological complexity of urbanisation, unless policy makers listen to scientists and scientific knowledge is rigorously applied in planning and construction.

Loss of life and damage may be greater anywhere in the world compared to similarly sized earthquakes in the past.

Another problem is that, in the future, tragic earthquakes may surprise us in unexpected places simply because the Earth's interior is not sufficiently known. In many parts of the world, attitude to the seismic hazard is often at odds with the actual level of threat.

Politicians, seismologists and engineers will have to work hard to predict as accurately as possible what will happen at a given site during a future earthquake, even if we don't know when it will happen.

P. Moczo et al., *Earthquakes*, Springer Praxis Books,
https://doi.org/10.1007/978-3-031-64707-9_12

Another, and fortunately rather pleasant and exciting, is the seismology of space exploration. Seismometers will be among the first instruments to be carried to large space bodies. Because of the potential extraction of mineral resources, because of the study of the evolution of our Solar System and space, and last but not least, because of safety on habitable planets.

Glossary

aftershock an earthquake that occurs as a consequence of the main earthquake (main shock) in the vicinity of the fault segment where the rupture has spread; the larger the earthquake, the more aftershocks (tens to thousands); the magnitude of the largest aftershock is on average about 1.2 units smaller than the magnitude of the earthquake

amplitude of a wave the largest displacement of a wave

asthenosphere mechanically weak part of the upper mantle located beneath the lithosphere; rocks within it can deform without macroscopic fracturing (macroscopic cracks and ruptures)

attenuation of wave decrease in wave amplitude as a function of distance from the source; decrease of a given wave can be caused by (a) material attenuation—internal friction due to the fact that the Earth's interior is not a perfectly elastic material, (b) scattering due to small inhomogeneities, (c) geometric spreading of the wave front, (d) splitting of the wave at material interfaces (the incident wave breaks down into reflected and refracted waves)

earthquake/tectonic earthquake the spontaneous initiation and propagation of a rupture along a fault, generating seismic waves by the propagating rupture, propagation of seismic waves inside the Earth, and seismic (vibrational) motion of the Earth's surface

focus of an earthquake the original term for the hypocentre

Earth's core the central part of the Earth's interior made up of iron and nickel and consisting of an outer liquid core (at depths approximately 2890–5155 km) and an inner solid core (5155–6378 km)

elastic material (or elastic solid) an ideal physical model of material behaviour in which the application of force or stress to a certain volume of material immediately

P. Moczo et al., *Earthquakes*, Springer Praxis Books,
https://doi.org/10.1007/978-3-031-64707-9

results in volume deformation, and upon removal of the force, the volume immediately returns to its original shape; for example, if an elastic rod is stretched, its elongation is instantaneous and directly proportional to the magnitude of the pull

epicentral intensity the value of macroseismic intensity at the epicentral area of an earthquake according to a chosen macroseismic scale; since macroseismic intensity characterizes the degree of earthquake effects on humans, objects, structures, and nature at a given location, epicentral intensity does not have to be the highest intensity in a given earthquake and does not have a direct relationship with the earthquake magnitude

epicentre it is the local radial projection (locally vertically upwards) of the hypocentre onto the Earth's surface

fault see tectonic fault

foreshock an earthquake that precedes a larger main earthquake (mainshock), with its hypocentre being relatively close to the hypocentre of the main earthquake; while aftershocks typically occur after every major earthquake, foreshocks do not occur in all active zones and/or for all mainshocks

free oscillations of the Earth the vibrations of the entire Earth resulting from the constructive interference of long-period surface seismic waves during and following major earthquakes; over 1 400 modes of free oscillations of the Earth have been identified

frequency a characteristic of a wave or vibrational motion, the inverse of the period; if a wave has a constant frequency, then, for example, at a frequency of 1 Hz (Hertz), the crest of the wave repeats itself every one second, at a frequency of 2 Hz, the crest of the wave repeats itself every half second

geophone a relatively robust and easily installable seismometer primarily used in seismic exploration to record artificially generated seismic waves; it records the velocity at which the soil vibrates

hypocentre the point of origin of an earthquake; the location on a fault where a rupture is initiated when stress exceeds the strength limit of the contact between two lithospheric plates (or blocks of the Earth's crust); the strength limit is determined by static friction and also influenced by the presence and pressure of fluids

liquefaction of sandy sediments loss of strength in water-saturated surface granular soils due to intense vibrational motion causing the compaction of loose granular deposits, resulting in turn in a rise of pore water pressure

lithosphere the mechanically rigid outer layer of the Earth composed of the crust and a portion of the upper mantle; the oceanic lithosphere has a thickness of approximately 100 km, while the thickness of the continental lithosphere ranges from approximately 100 to 250 km; the lithosphere is broken into lithospheric plates that are in motion relative to each other; the largest plates include the Pacific, North American, Eurasian, African, Antarctic, Indo-Australian (sometimes Australian and Indian plates are distinguished), and South American plates

lithospheric plate see lithosphere

Love wave one of the two basic types of surface seismic waves; particles oscillate horizontally perpendicular to the direction of wave propagation; the existence of the wave depends on the Earth's surface and the speed of shear waves beneath the surface, which is lower than at greater depths; it results from the superposition (interference) of body waves, and as a consequence, there are multiple modes of Love waves, each characterized by a different velocity-frequency dependence; the amplitude of the mode at a given frequency decreases exponentially with depth beneath the surface layer (surface layers)

M the symbol used for an unspecified type of earthquake magnitude

macroseismic intensity a quantity that characterizes and quantifies the effects of an earthquake on people, objects, structures, and nature in degrees of the macroseismic scale; the value of macroseismic intensity is determined for each location affected by macroseismic effects, with these effects described in macroseismic questionnaires by direct observers

macroseismic scale a scale used to characterize the effects of an earthquake on people, objects, structures, and nature at a given location; in Europe, the most commonly used scale is the 12-degree European Macroseismic Scale (EMS-98); degree I corresponds to a situation where no one felt the earthquake; degree XII signifies the destruction of most building structures and changes in the landscape

magnitude a quantity that quantifies the size of an earthquake; various types of original magnitudes based on the idea of Japanese seismologist Kiyoo Wadati and American seismologist Charles Francis Richter have a fundamental problem— beyond a certain earthquake size, they do not allow distinguishing between two different large earthquakes; this is because they do not sufficiently consider the physics of fault rupture propagation during earthquakes

moment magnitude a quantity that quantifies the size of an earthquake; its value is derived from the seismic moment; this magnitude does not suffer from saturation like previously developed magnitudes and effectively distinguishes the size of even the largest earthquakes because it reasonably accounts for rupture propagation on the fault

Ms a symbol for earthquake magnitude determined from records of surface seismic waves; it belongs to magnitudes developed before the introduction of the moment magnitude

Mw a symbol for the moment magnitude

P-wave longitudinal (compressional) seismic wave; as the P-wave passes through, particles vibrate in the direction of wave propagation, causing only volume changes; P-wave propagates at least 1.41 times faster than the S-wave in any medium

particle acceleration the acceleration at which a particle vibrates; maximum particle acceleration (often called peak ground acceleration) is one of the important characteristics of seismic motion at a given location

particle displacement the displacement of a particle during its vibrational motion; maximum particle displacement is one of the important characteristics of seismic motion at a given location

particle velocity the velocity at which a particle vibrates; maximum particle velocity is one of the important characteristics of seismic motion at a given location

period a characteristic of a wave or vibrational motion; the inverse of frequency; if a wave has a constant period, at a period of 1 second, the crest of the wave repeats every one second, and at a period of 2 seconds, the crest of the wave repeats every two seconds

Rayleigh wave the more complex of the two fundamental surface seismic waves; particles on the Earth's surface undergo an elliptical motion which is most often retrograde but may be prograde in case of very soft surface material; the existence of the wave depends on the Earth's surface; it results from the superposition (interference) of body waves, and as a consequence, there are multiple modes of Rayleigh waves, each characterized by a different velocity-frequency dependence; the amplitude of the mode at a given frequency decreases exponentially with depth beneath the surface layer (surface layers)

resonant frequency at a site the frequency at which seismic motion is significantly (resonantly) amplified by the constructive interference of waves due to the transmission properties of the underground site conditions; in the case of an ideal horizontal sedimentary layer overlying bedrock, seismic motion is amplified at a frequency corresponding to a wavelength equal to four times the thickness of the layer

rheology the branch of physics that studies the relationship between the deformation of a material and the forces/stresses that cause the deformation; "rheology of the Earth's interior" succinctly refers to the relationship between deformation and stresses for materials inside the Earth

rupture a temporary loss of contact or loss of strength of contact at a specific location in a material, or a loss of contact or strength of the original contact along a fault; if two material particles were adjacent before the formation of the rupture, the formation of the rupture causes them to move apart—along a fault without or with creating a void

S-wave shear body seismic wave; as the S-wave passes through, particles vibrate in a direction perpendicular to the direction of wave propagation, causing only shape changes; S-wave cannot propagate in fluids; in any solid medium, the S-wave propagates at least 1.41 times slower than the P-wave

scattering of waves a wave phenomenon in which the direction of wave propagation is altered by relatively small inhomogeneities in the medium; intense scattering can occur in "broken," more or less fragmented or fractured rocks; significant scattering of seismic waves in the fragmented crust of the Moon, along with relatively weak wave attenuation, is the cause of the very long duration of moonquake recordings

seismic model of the Earth a model of the spatial distribution of material parameters within the Earth based on the propagation of seismic waves in the Earth, which gives the speeds of P and S-waves, densities, and quality factors of P and S-waves (see also wave attenuation)

seismic moment a quantity that quantifies the size of an earthquake; its physical unit is Nm (Newton times metre); it is equal to the product of the average shear modulus, the area of the ruptured part of the fault (the area where the rupture from the hypocentre has spread), and the average slip (offset) on the ruptured part of the fault

seismic noise (seismic ambient noise) continuous very weak seismic motion that is visible on seismic records even in the absence of transient seismic events, such as earthquakes or explosions; it is contributed to by thousands of natural (sea waves, wind, ...) or anthropic (traffic, machinery, ...) sources of mechanical vibration, most of which cannot even be identified; seismic noise essentially forms the seismic background of any transient phenomenon that we would like to analyse without it being disturbed by seismic noise

seismic waves the propagation of mechanical vibrational motion within the Earth; the four basic types of seismic waves in the Earth are the body P and S-waves, and the surface Love and Rayleigh waves

seismogram a record of the particle displacement, particle velocity, or particle acceleration of seismic (vibrational) motion at a given location; it can be a record from a seismograph or a numerically simulated "record."

seismograph a device for recording seismic motion as a function of time

seismometer the part of a seismograph that senses the particle displacement, particle velocity, or particle acceleration of seismic (vibrational) motion at a given location on Earth

site effect of an earthquake an occurrence of anomalous seismic motion (e.g., large amplitudes, prolonged duration) or macroseismic effect that is spatially localized and significantly differs from the motions and effects in the surrounding area

shock a term used for an earthquake; the main shock refers to the earthquake itself, followed by smaller earthquakes called aftershocks

speed of wave propagation a property of a material (e.g., rock) that characterizes how fast a body P-wave or body S-wave can propagate; in the case of surface waves, which have an interference nature (formed by the superposition of body waves), it is important to distinguish between phase velocity (the velocity at which the phase of the wave propagates) and group velocity (the velocity at which the wave's energy maximum propagates; in general smaller than the phase velocity)

stress the force per unit area exerted across a surface; it should be distinguished from volume non-contact force (e.g., gravitational force); when contact forces act on a volume (through surface of an object or through a surface of just a hypothetical volume within the object), they create a state of stress inside the volume; imagine that if some surface within the volume divides the volume into part A and part B, the force contact action of part A on part B at a given point on the surface is characterized by the stress vector; for readers with a background in physics it should be noted that the stress state at a given point is fully characterized by the six values of the stress tensor components

subduction the process of one lithospheric plate sliding or diving beneath another; a good example is the subduction of the Nazca Plate beneath the South American Plate along the western coast of South America

supershear rupture propagation the propagation of a rupture along a fault at a velocity greater than the speed of S-waves; one could loosely analogize it to supersonic propagation in the air

tangential stress on a fault a component of stress vector parallel to the fault plane, which characterizes how one plate "pulls" in the direction of its motion the other plate

tectonic fault a zone of contact between two lithospheric plates or blocks of the Earth's crust where there has been or is ongoing relative motion; at most plate boundaries, continuous and smooth motion is impeded by static friction, preventing free "sliding"; see also earthquake

tsunami a long and fast ocean wave that involves the entire water layer from the surface to the ocean floor; tsunami is generated by a sudden uplift of the sea floor (most commonly due to earthquakes in subduction zones) or a subsidence (due to a big landslide) or a volcano eruption; in the open ocean, it can propagate at velocities of approximately 500–900 km/h; with wavelengths of several hundred kilometres and amplitudes not exceeding several tens of centimetres, it is unnoticeable and harmless in the open ocean; it becomes dangerous near coastlines, where the reduction in velocity, and thus wavelength, causes a significant increase in the amplitude (maximum wave height)

viscoelastic solid a material that exhibits elastic (spring-like) behaviour under short-term force application at a given temperature and pressure, and viscous behaviour under long-term force application

viscous liquid a liquid that exhibits resistance to deformation under shear stresses, or in simpler terms, has a certain degree of "reluctance" to flow; one can recall the "reluctance" of honey to flow compared to water, which has significantly lower viscosity and therefore spreads easily and quickly

wavelength the distance between two wave crests; it is equal to the product of the period and the velocity of wave propagation

References

Agnew, D. C. (2002). History of seismology. In W. H. K. Lee, P. Jennings, C. Kisslinger, & H. Kanamori (Eds.), *International handbook of earthquake and engineering seismology* (Vol. B1A, pp. 3–11). Academic Press.

Aguirre, B. E. (2018). A retrospective account of the impacts of the 1960 Valdivia, Chile, earthquake and tsunami and the lack of business continuity planning. In K. J. Engemann (Ed.), *The Routledge companion to risk, crisis and security in business* (pp. 93–99). Routledge.

Ambraseys, N. N. (1989). Temporary seismic quiescence: SE Turkey. *Geophysical Jou rnal International, 96,* 311–331.

Ambraseys, N. N. (2001). The earthquake of 1509 in the Sea of Marmara, Turkey, revisited. *Bulletin of the Seismological Society of America, 91,* 1397–1416.

Ammon, C. J., Chen, J., Hong-Kie, T., et al. (2005). Rupture process of the 2004 Sumatra-Andaman earthquake. *Science, 308,* 1133–1139.

Amundsen, M. A. (2017). Seeing Arizona, imagining mars. Deserts, canals, global climate change, and the American West. *The Journal of Arizona History, 58,* 331–350.

Andreoni, A. (2014). Mythology and earthquakes in Italian literature of the 18th century. *Forum Italicum, 48,* 126–131.

Angell, E. (2012). A seismic cityscape: Earthquakes in Istanbul's history. In Aydin, M. A. (Ed.) *History of Istanbul from antiquity to XXI century, Vol. 1.*

Archuleta, R. J. (1984). A faulting model for the 1979 Imperial Valley earthquake. *Journal of Geophysical Research, 89,* 4559–4585.

Armijo, R., Meyer, B., Barka, A., et al. (2000). The fault breaks of the 1999 earth-quakes in Turkey and the tectonic evolution of the Sea of Marmara: A summary.

In A. Barka, O. Kozaci, S. Akyüz, & E. Altunel (Eds.), *The 1999 Izmit and Düzce earthquakes: Preliminary results* (pp. 55–62). Istanbul Technical University.

Armijo, R., Meyer, B., Hubert, A., et al. (1999). Westward propagation of the North Anatolian Fault into the northern Aegean: Timing and kinematics. *Geology, 27*, 267–270.

Astiz, L., Kanamori, H., & Eissler, H. (1987). Source characteristics of earthquakes in the Michoacan seismic gap in Mexico. *Bulletin of the Seismological Society of America, 77*, 1326–1346.

Atwater, B. F., Cisternas, V. M., Bourgeois, J. et al. (2005). *Surviving a tsunami – lessons from Chile, Hawaii and Japan*. Circular 1187, Version 1.1, Revised Edition. U.S. Geological Survey.

Bard, P.-Y., & Bouchon, M. (1985). The two-dimensional resonance of sediment-filled valleys. *Bulletin of the Seismological Society of America, 75*, 519.

Barka, A. (1999). The 17 August 1999 Izmit Earthquake. *Science, 285*, 1858–1859.

Bath, M. (1973). *Introduction to Seismology*. Birkhauser Verlag.

Beckmann, M. (2011). *The column of Marcus Aurelius*. The University of North Carolina Press.

Benioff, H., Press, F., & Smith, S. (1961). Excitation of the free oscillations of the Earth by earthquakes. *Journal of Geophysical Research, 66*, 605–619.

Ben-Menahem, A. (1995). A Concise history of mainstream seismology: Origins, legacy, and perspectives. *Bulletin of the Seismological Society of America, 85*, 1202–1225.

Berberian, M. (2014). Earthquake myths. In Shroder, J.F. (Ed.) *Developments in earth surface processes, 17*, 23–41.

Bolt, B. A. (1982). *Inside the Earth: Evidence from earthquakes*. W.H. Freeman.

Bolt, B. A. (1993). *Earthquakes and geological discovery*. Scientific American Library.

Boschi, E., Caserta, A., Conti, C., et al. (1995). Resonance of subsurface sediments: an unforeseen complication for designers of Roman columns. *Bulletin of the Seismological Society of America, 85*, 320–324.

Bosworth, R. J. B. (1981). The Messina earthquake of 28 December 1908. *European Studies Review, 11*, 189–206.

Bouchon, M., & Karabulut, H. (2008). The aftershock signature of supershear earthquakes. *Science, 320*, 1323–1325.

Bouchon, M., Bouin, M. P., Karabulut, H., et al. (2001). How fast is rupture during an earthquake? New insights from the 1999 Turkey earthquakes. *Geophysical Research Letters, 28*, 2723–2726.

Bouchon, M., Karabulut, H., Bouin, M.-P., et al. (2010). Faulting characteristics of supershear earthquakes. *Tectonophysics, 493*, 244–253.

Bouchon, M., Karabulut, H., Aktar, M., et al. (2011). Extended nucleation of the 1999 Mw 7.6 Izmit earthquake. *Science, 331*, 877–880.

Bouchon, M., Karabulut, H., Aktar, M., et al. (2021). The nucleation of the Izmit and Düzce earthquakes: Some mechanical logic on where and how ruptures began. *Geophysical Journal International, 225*, 1510–1517.

Bouin, M.-P., Bouchon, M., Karabulut, M., & Aktar, M. (2004). Rupture process of the 12 November 1999, Düzce (Turkey) earthquake deduced from strong motion and GPS measurements. *Geophysical Journal International, 159*, 207–211.

Bozkurt, E. (2001). Neotectonics of Turkey – a synthesis. *Geodinamica Acta, 14*, 3–30.

Campillo, M., Gariel, J. C., Aki, K., & Sánchez-Sesma, F. J. (1989). Destructive strong ground motion in Mexico City: source, path, and site effects during great 1985 Michoacan earthquake. *Bulletin of the Seismological Society of America, 79*, 1718–1735.

Cassius Dio. Roman history, LXVIII, 24–25

Cholnoky, J. (1906). Földrajzi közlemények. *Bulletin de la Société Hongroise de géographie XXXIV*

Choy, G. L., & Boatwright, J. (2007). The energy radiated by the 26 December 2004 Sumatra–Andaman earthquake estimated from 10-minute P-wave windows. *Bulletin of the Seismological Society of America, 97*, S18–S24.

Coen, D. R. (2013). *The earthquake observers. Disaster science from Lisbon to Richter.* The University of Chicago Press.

Collier, S., & Sater, W. F. (2004). *A history of Chile, 1808-2002.* Cambridge University Press.

Cruz-Atienza, V. M., Tago, J., Sanabria-Gómez, J. D., et al. (2016). Long duration of ground motion in the paradigmatic Valley of Mexico. *Nature Scientific Reports, 6*, 38807.

Davison, C. (1921a). Founders of Seismology—I. John Michell. *Geological Magazine, 58*, 98–107.

Davison, C. (1921b). Founders of Seismology.—II. Robert Mallet. *Geological Magazine, 58*, 241–250.

Davison, C. (1921c). Founders of seismology.—III. John Milne. *Geological Magazine, 58*, 385–396.

Davison, C. (1937). Founders of seismology, IV. *Geological Magazine, 74*, 529–534.

Defra Flood Management Division, the Health and Safety Executive and the Geological Survey of Ireland 2006. Tsunamis – Assessing the hazard for the UK and Irish coasts

Dewey, J., & Byerly, P. (1969). The early history of seismometry (to 1900). *Bulletin of the Seismological Society of America, 59*, 183–227.

Dewey, J. W., Choy, G., Presgrave, B., et al. (2007). Seismicity associated with the Sumatra–Andaman Islands earthquake of 26 December 2004. *Bulletin of the Seismological Society of America, 97*, S25–S42.

Drilleau, M., Samuel, H., & Garcia, R. F. (2022). Marsquake locations and 1-D seismic models for mars from InSight data. *Journal of Geophysical Research: Planets, 127*.

Duman, T. Y., & Emre, Ö. (2013). *The East Anatolian Fault: Geometry, segmentation and jog characteristics.* Geological Society, London, Special Publications, SP372.14.

Dynes, R. R. (1999). *The dialogue between voltaire and rousseau on the Lisbon earthquake: The emergence of a social science view.* University of Delaware, Disaster Research Center.

Dziewonski, A., & Romanowicz, B. A. (2007). Overview. In A. Dziewonski & B. A. Romanowicz (Eds.), *Treatise on geophysics* (Seismology and the structure of the Earth) (Vol. 1, pp. 1–29). Elsevier.

Ebel, J. E. (2006). The Cape Ann, Massachusetts earthquake of 1755: A 250th anniversary perspective. *Seismological Research Letters, 77,* 74–86.

Erşan, S. (2016). A comparative evaluation of the results of two earthquakes: Istanbul and Lisbon earthquake in 18th Century. In H. Cruz, J. Saporiti Machado, A. Campos Costa, et al. (Eds.), *Historical earthquake-resistant timber framing in the Mediterranean area* (Vol. 1, pp. 47–55). LNCE.

Esteva, L. (1988). Consequences, lessons, and impact on research and practice. *Earthquake Spectra, 4,* 413–426.

Farrell, E. J., Ellis, J. T., & Hickey, K. R. (2015). Tsunami case studies. In *Coastal and marine hazards, risks, and disasters.* Elsevier.

Ferrari, G., & McConnel, A. (2005). Robert Mallet and the 'Great Neapolitan earthquake' of 1857. *Notes and Records: the Royal Society, 59,* 45–64.

Filizzola, C., Corrado, A., Genzano, N., et al. (2022). RST analysis of anomalous TIR sequences in relation with earthquakes occurred in Turkey in the period 2004–2015. *Remote Sensing, 14,* 381.

Fradkin, P. L. (2001). *Wildest Alaska: journeys of great peril in Lituya Bay.* University of California Press.

Frendo, J. D. (Ed.). (1975). *Agathias scholasticus. The histories.* Walter de Gruyter & Co.

Friedrich, W. L. (2013). The Minoan Eruption of Santorini around 1613 BC and its consequences. *Tagungen des Landesmuseums für Vorgeschichte Halle, 9,* 37–48.

Frohlich, C., & Nakamura, Y. (2009). The physical mechanisms of deep moonquakes and intermediate-depth earthquakes: How similar and how different? *Physics of the Earth and Planetary Interiors, 173,* 365–374.

Funiciello, R., & Rovelli, A. (1998). Terremoti e monumenti in Roma. *Le Scienze, 357,* 42–49.

Garcia, R. F., Khan, A., Drilleau, M., et al. (2019). Lunar seismology: An update on interior structure models. *Space Science Reviews, 215*(8).

Giardini, D., Lognonné, P., Banerdt, W. B., et al. (2020). The seismicity of Mars. *Nat Geoscience, 13,* 205–212.

Guidoboni, E., Comastri, A., & Traina, G. (1994). *Catalogue of ancient earthquakes in the Mediterranean area up to the 10th century.* Instituto Nazionale di Geofisica.

Güvercin, S. E., Karabulut, H., Özgün Konca, A., et al. (2022). Active seismotectonics of the East Anatolian Fault. *Geophysical Journal International, 230,* 50–69.

Hamnett, B. (2004). *A concise history of Mexico.* Cambridge University Press.

Hertz, P. (2014). The present and future of space science at NASA. *Proceedings of the American Philosophical Society, 158,* 329–353.

Hough, S. E., & Bilham, R. G. (2006). *After the Earth quakes: Elastic rebound on an urban planet.* Oxford University Press.

Hough, E., Bilham, R. G., Ambraseys, N., & Feldl, N. (2005). Revisiting the 1897 Shillong and 1905 Kangra earthquakes in northern India: Site response, Moho reflections and a triggered earthquake. *Current Science, 88*, 1632–1638.

Ide, S., Baltay, A., & Beroza, G. C. (2011). Shallow dynamic overshoot and energetic deep rupture in the 2011 Mw 9.0 Tohoku-Oki earthquake. *Science, 332*, 1426–1429.

International Atomic Energy Agency (2015). The Fukushima Daiichi Accident. *Report by the Director General.*

Jackson, J. (2006). Fatal attraction: living with earthquakes, the growth of villages into megacities, and earthquake vulnerability in the modern world. *Philosophical Transactions of the Royal Society, 364*, 1911–1925.

Jeffreys, E., Jeffreys, M., & Scott, R. (Eds.). (2017). *The chronicle of John Malalas.* Byzantina Australiensia.

Johnston, D., et al. (2008). Developing an effective tsunami warning system: Lessons from the 1960 Chile earthquake tsunami for New Zealand coastal communities. *Kotuitui: New Zealand Journal of Social Sciences Online, 3*, 105–120.

Kanamori, H., & Anderson, L. (1975). Amplitude of the Earth's free oscillations and long-period characteristics of the earthquake source. *Journal of Geophysical Research, 80*, 1075–1078.

Kanamori, H., & Cipar, J. J. (1974). Focal process of the Great Chilean earthquake May 22, 1960. *Physics of the Earth and Planetary Interiors, 9*, 128–136.

Karabulut, H., Güvercin, S. E., Hollingsworth, J., et al. (2023). Long silence on the East Anatolian fault zone (Southern Turkey) ends with devastating double earthquakes (6 February 2023) over a seismic gap: implications for the seismic potential in the Eastern Mediterranean region. *Journal of the Geological Society, 180*, jgs2023-021.

Kerr, R. A. (2011). A quake may have hinted that it was on the way. *Science, 331*, 836.

Khan, K., Ceylan, S., van Driel, M., et al. (2021). Upper mantle structure of Mars from InSight seismic data. *Science, 373*, 434–438.

Kloeg, P. (2013). *Antioch the Great: Population and economy of second-century Antioch.* Dissertation.

Knapmeyer-Endrun, B., Panning, M. P., Bissig, F., et al. (2021). Thickness and structure of the martian crust from InSight seismic data. *Science, 373*, 438–443.

Knost, S. (2010). The impact of the 1822 earthquake on the administration of waqf in Aleppo. In P. Sluglett & S. Weber (Eds.), *Syria and Bilad al-Sham under Ottoman rule: Essays in honour of Abdul-Karim Rafeq* (pp. 293–309).

Knost, S. (2017). Living with Disaster: Aleppo and the earthquake of 1822. In G. J. Schenk (Ed.), *Historical disaster experiences* (pp. 295–305). Springer International Publishing.

LaMoreaux, P. E. (1995). Worldwide environmental impacts from the eruption of Thera. *Environmental Geology, 26*, 172–181.

Lane, K. M. D. (2011). *Geographies of Mars: Seeing and knowing the red planet.* The University of Chicago Press.

Larsen, S. E. (2006). The Lisbon earthquake and the scientific turn in Kant's philosophy. *European Review, 14*, 359–367.

Lawson, A. C. (chairman) (1908). *The California earthquake of April 18, 1906: Report of the State Earthquake Investigation Commission, vol. I*. Carnegie Institution of Washington Publication 87

Lawson, A. C., & Byerly, P. (1951). *Harry Fielding Reid 1859-1944*. National Academy of Sciences.

Lillie, R. J. (1998). *Whole Earth Geophysics: An introductory textbook for geologists and geophysicists*. Pearson.

Lognonné, P., & Pike, W. T. (2015). Planetary seismometry. In V. C. H. Tong & R. A. García (Eds.), *Extraterrestrial seismology* (pp. 36–48). Cambridge University Press.

Losey, R. J. (2007). Native American vulnerability and resiliency to great Cascadia earthquakes. *Oregon Historical Quarterly, 108*, 201–221.

Lowrie, W. (2018). *Geophysics: A very short introduction*. Oxford University Press.

Ludwin, R. S., Dennis, R., Carver, D., et al. (2005). Dating the 1700 Cascadia earthquake: Great coastal earthquakes in Native Stories. *Seismological Research Letters, 76*, 140–148.

Maas, M. (2005). *The Cambridge companion to the age of Justinian*. Cambridge University Press.

Mai, P. M., Aspiotis, T., Aquib, T. A., et al. (2023). The destructive earthquake doublet of 6 February 2023 in South-central Türkiye and Northwestern Syria: Initial observations and analyses. *The Seismic Record, 3*(2), 105–115.

Matias, L. (2015). *Possible sources of the 1755 earthquake. Presentation*. Universidade de Lisboa.

Matouš, L. (Ed.) (1958). *Epos o Gilgamešovi. Státní nakladatelství krásné literatury, hudby a umění*.

Meier, M. (2006). Natural disasters in chronographia of John Malalas: Reflections on their function – An initial sketch. *The Medieval History Journal, 10*, 237–266.

Melgar, D., Taymaz, T., Ganas, A., et al. (2023). Sub- and super-shear ruptures during the 2023 Mw 7.8 and Mw 7.6 earthquake doublet in SE Türkiye. *Seismica, 2*(3).

Miller, D. J. (1960). *Giant waves in Lituya Bay Alaska. Shorter contributions to general geology*. USGS.

Minoura, K., Imamura, F., Sugawara, D., et al. (2001). The 869 Jōgan tsunami deposit and recurrence interval of large-scale tsunami on the Pacific coast of northeast Japan. *Journal of Natural Disaster Science, 23*, 83–88.

Moczo, P., Rovelli, A., Labák, P., & Malagnini, L. (1995). Seismic response of the geologic structure underlying the Roman Colosseum and 2-D resonance of a sediment valley. *Annali di Geofisica, 38*, 939–956.

Moczo, P., Kristek, J., & Gális, M. (2014). *The finite-difference modelling of earthquake motions. Waves and ruptures*. Cambridge University Press.

Moczo, P., Rutšeková, E., Kristek, J., Gális, M., & Kristeková, M. (2023). *Zemetrasenia. Tragické výzvy v dejinách*. Grada.

Molesky, M. (2016). *This Gulf of Fire*. Vintage Books.

Moreno, J. M. (1995). The 1985 Mexico earthquake. *Geofísica Colombiana, 3*, 5–19.

Moss, E., & Moffat, R. (2020). Prior and future earthquake effects in Valdivia, Chile. *Obras y Proyectos, 27*, 41–49.

Nakamura, Y. (2015). Planetary seismology: Early observational results. In V. C. H. Tong & R. A. García (Eds.), *Extraterrestrial seismology* (pp. 91–106). Cambridge University Press.

Nakamura, Y. (2020). Rebirth of extraterrestrial seismology. *Nature Geoscience, 13*, 178–179.

Needham, J. (1959). *Science and civilisation in China* (Vol. 3). Cambridge University Press.

Nunn, C. R. F., Garcia, Y. N., et al. (2020). Lunar seismology: A data and instrumentation review. *Space Science Reviews, 216*, 89.

Oldham, R. D. (1899). Report on the Great earthquake of 12th June 1897. *Memoirs of the Geological Survey of India, XXXIX*.

Palutoğlu, M., & Şaşmaz, A. (2017). 29 November 1795 Kahramanmaraş earthquake, southern Turkey. *Bulletin of the Mineral Research and Exploration, 155*, 187–202.

Park, J., Song, T.-R. A., Tromp, J., et al. (2005). Earth's free oscillations excited by the 26 December 2004 Sumatra-Andaman earthquake. *Science, 308*, 1139.

Přidal, A. (2011). *Potulky knihami a časem*. Barrister & Principal.

Rabinovich, A. B., Titov, V. V., Moore, C. W., et al. (2017). The 2004 Sumatra tsunami in the Southeastern Pacific Ocean: New global insight from observations and modeling. *Journal of Geophysical Research: Oceans, 122*, 7992–8019.

Rajendran, N. (2006). History of Tsunami. In S. M. Ramasamy, C. J. Kumanan, B. R. Sivakumar, & B. Singh (Eds.), *Geomatics in tsunami* (pp. 1–9). New India Publishing Agency.

Rector, J. L. (2003). *The history of Chile*. Palgrave Macmillan.

Reid, H. F. (1910). The mechanics of the earthquake, vol. II of Lawson, A.C. (chairman). *The California earthquake of April 18, 1906: Report of the State Earthquake Investigation Commission*. Carnegie Institution of Washington Publication 87.

Robinson, A. (2012). *Earthquake: Nature and culture*. Reaktion Books Ltd.

Robinson, A. (2016). *Earth-shattering events*. Thames & Hudson Inc..

Rosakis, A., Abdelmeguid, M., & Elbanna, A. (2023). *Evidence of early supershear transition in the Mw 7.8 Kahramanmaraş earthquake from near-field records*. arXiv:2302.07214 [physics.geo-ph].

Rousseau, J. J. (1992). Letter to voltaire. In R. D. Masters & C. Kelly (Eds.), *The collected writings of Rousseau* (Vol. 3, pp. 49–52). University Press of New England.

Samuel, H., et al. (2023). Geophysical evidence for an enriched molten silicate layer above Mars's core. *Nature, 622*, 712–717.

Severn, R. T. (2012). Understanding earthquakes: From myth to science. *The Bulletin of Earthquake Engineering, 10*, 351–366.

Sezen, H., et al. (2000). *Structural engineering reconnaissance of the August 17, 1999 Earthquake: Kocaeli (Izmit), Turkey* (pp. 9–29). Pacific Earthquake Engineering Research Center, University of California.

Shearer, P. (2019). *Introduction to seismology* (3rd ed.). Cambridge University Press.

Silva, M. G. (2003). Discourse on earthquakes. *Revista de Ingeniería, 18*, 71–84.

Simons, M., Minson, S. E., Sladen, A., et al. (2011). The 2011 magnitude 9.0 Tohoku-Oki earthquake: Mosaicking the megathrust from seconds to centuries. *Science, 332*, 1421–1425.

Stähler, S. C., Khan, A., Banerdt, W. B., et al. (2021). Seismic detection of the Martian core. *Science, 373*, 443–448.

Stähler, S. C., Mittelholz, A., Perrin, C., et al. (2022). Tectonics of Cerberus Fossae unveiled by marsquakes. *Nature Astronomy, 6*, 1376–1386.

Stein, S., & Wysession, M. (2013). *An introduction to seismology, earthquakes, and Earth structure*. Wiley-Blackwell.

Stein, R. S., Barka, A. A., & Dieterich, J. H. (1997). Progressive failure on the North Anatolian fault since 1939 by earthquake stress triggering. *Geophysical Journal International, 128*, 594–604.

Stone, W. C., Yokel, F. Y., Celebi, M., et al. (1987). Engineering aspects of the September 19, 1985 Mexico earthquake. In *NBS Building Science Series 165*. U.S. Department of Commerce.

Tanimoto, T., & Anderson, A. (2023). Seismic noise between 0.003 Hz and 1.0 Hz and its classification. *Progress in Earth and Planetary Sciences, 10*, 56.

Tunc, T. E., & Tunc, G. (2022). Transferring technical knowledge to Turkey: American Engineers, Scientific Experts, and the Erzincan Earthquake of 1939. *Notes and Records, 76*, 387–406.

Voltaire. (2013). *Poem on the Lisbon Disaster: Or an inquiry into the axiom, "All is well"* (Lyon, A., Trans.)

Weber, R. C., & Knapmeyer, M. (2015). Seismicity and interior structure of the Moon. In V. C. H. Tong & R. A. García (Eds.), *Extraterrestrial seismology* (pp. 203–224). Cambridge University Press.

Weber, R. C., Knapmeyer, M., Panning, M., et al. (2015). Modeling approaches in planetary seismology. In V. C. H. Tong & R. A. García (Eds.), *Extraterrestrial seismology* (pp. 140–155). Cambridge University Press.

Website of the National Aeronautics and Space Administration (NASA), Public domain.

Website of the National Oceanic and Atmospheric Administration (NOAA), Public domain.

Website of the United States Geological Survey (USGS), Public domain.

Williams, R. A., McCallister, N. S., & Dart, R. L. (2011). *20 cool facts about the New Madrid seismic zone—Commemorating the bicentennial of the New Madrid earthquake sequence, December 1811–February 1812 (poster): USGS General Information Product 134*.

Yeats, R. (2015). *Earthquake time bombs*. Cambridge University Press.

Zilio, L. D., & Ampuero, J.-P. (2023). Earthquake doublet in Turkey and Syria. *Communications Earth and Environment, 4*, 71.

Zoback, M.L. (2006). The 1906 earthquake and a century of progress in understanding earthquakes and their hazards. *GSA Today*, April/May.